FEDERAL EXECUTIVE TEAM

Director, U.S. Climate Change Science Program..William J. Brennan

Director, Climate Change Science Program Office...Peter A. Schultz

Lead Agency Principal Representative to CCSP;
Associate Director, Department of Energy, Office of Biological
and Environmental Research ....Anna Palmisano

Product Lead; Department of Energy,
Office of Biological and Environmental Research ....Anjuli S. Bamzai

Synthesis and Assessment Product Advisory Group Chair;
Associate Director, EPA National Center for Environmental
Assessment...Michael W. Slimak

Synthesis and Assessment Product Coordinator,
Climate Change Science Program Office ....Fabien J.G. Laurier

OTHER AGENCY REPRESENTATIVES

National Aeronautics and Space Administration ....Donald Anderson
National Oceanic and Atmospheric Administration ....Brian D. Gross
National Science Foundation ....Jay S. Fein

Climate Models

An Assessment of Strengths and Limitations

Synthesis and Assessment Product 3.1
Report by the U.S. Climate Change Science Program
and the Subcommittee on Global Change Research

AUTHORS:

David C. Bader, Lawrence Livermore National Laboratory
Curt Covey, Lawrence Livermore National Laboratory
William J. Gutowski Jr., Iowa State University
Isaac M. Held, NOAA Geophysical Fluid Dynamics Laboratory
Kenneth E. Kunkel, Illinois State Water Survey
Ronald L. Miller, NASA Goddard Institute for Space Studies
Robin T. Tokmakian, Naval Postgraduate School
Minghua H. Zhang, State University of New York Stony Brook

CLIMATE CHANGE SCIENCE PROGRAM PRODUCT DEVELOPMENT ADVISORY COMMITTEE

Eight members of the Climate Change Science Progam Product Development Advisory Committee (CPDAC) wrote this Climate Change Science Program Synthesis and Assessment Product at the request of the Department of Energy. The entire CPDAC has accepted the contents of the product. Recommendations made in this report regarding programmatic and organizational changes, and the adequacy of current budgets, reflect the judgment of the report's authors and the CPDAC and are not necessarly the views of the U.S. Government.

Chair
Soroosh Sorooshian, University of California, Irvine

Vice Chair
Antonio J. Busalacchi, University of Maryland

Designated Federal Officer
Anjuli S. Bamzai, Department of Energy
Office of Biological and Environmental Research

Members

*David C. Bader
Lawrence Livermore National Laboratory

Virginia R. Burkett
U.S. Geological Survey

Leon E. Clarke
Pacific Northwest National Laboratory

*Curt Covey
Lawrence Livermore National Laboratory

James A. Edmonds
Pacific Northwest National Laboratory

Karen Fisher-Vanden
Dartmouth College

Brian P. Flannery
Exxon-Mobil Corporation

*William J. Gutowski Jr.
Iowa State University

David G. Hawkins
Natural Resources Defense Council

*Isaac M. Held
Geophysical Fluid Dynamics Laboratory

Henry D. Jacoby
Massachusetts Institute of Technology

David W. Keith
University of Calgary

*Kenneth E. Kunkel
Illinois State Water Survey

Richard S. Lindzen
Massachusetts Institute of Technology

Linda O. Mearns
National Center for Atmospheric Research

*Ronald L. Miller
National Aeronautics and Space Administration

Edward A. Parson
University of Michigan

Hugh M. Pitcher
Pacific Northwest National Laboratory

William A. Pizer
Resources for the Future

John M. Reilly
Massachusetts Institute of Technology

Richard G. Richels
Electric Power Research Institute

Cynthia E. Rosenzweig
National Aeronautics and Space Administration

*Robin T. Tokmakian
Naval Postgraduate School

Mort D. Webster
University of North Carolina

Julie A. Winkler
Michigan State University

Gary W. Yohe
Wesleyan University

*Minghua H. Zhang
State University of New York Stony Brook

*Authors for *Climate Models: An Assessment of Strengths and Limitations*

July 2008

Members of Congress:

On behalf of the National Science and Technology Council, the U.S. Climate Change Science Program (CCSP) is pleased to transmit to the President and the Congress this Synthesis and Assessment Product (SAP), *Climate Models: An Assessment of Strengths and Limitations*. This is part of a series of 21 SAPs produced by the CCSP aimed at providing current assessments of climate change science to inform public debate, policy, and operational decisions. These reports are also intended to help the CCSP develop future program research priorities.

The CCSP's guiding vision is to provide the Nation and the global community with the science-based knowledge needed to manage the risks and capture the opportunities associated with climate and related environmental changes. The SAPs are important steps toward achieving that vision and help to translate the CCSP's extensive observational and research database into informational tools that directly address key questions being asked of the research community.

This SAP assesses the strengths and limitations of climate models. It was developed with broad scientific input and in accordance with the Guidelines for Producing CCSP SAPs, the Federal Advisory Committee Act, the Information Quality Act (Section 515 of the Treasury and General Government Appropriations Act for Fiscal Year 2001 - Public Law 106-554), and the guidelines issued by the Department of Energy pursuant to Section 515.

We commend the report's authors for both the thorough nature of their work and their adherence to an inclusive review process.

Sincerely,

Carlos M. Gutierrez
Secretary of Commerce

Vice-Chair, Committee on
Climate Change
Science and Technology Integration

Samuel W. Bodman
Secretary of Energy

Chair, Committee on
Climate Change
Science and Technology Integration

John H. Marburger, III, Ph.D.
Director, Office of
Science and Technology Policy
Executive Director, Committee on
Climate Change
Science and Technology Integration

ACKNOWLEDGEMENT

This report has been peer reviewed in draft form by individuals chosen for their diverse perspectives and technical expertise. The expert review and selection of reviewers followed the OMB's Information Quality Bulletin for Peer Review. The purpose of this independent review is to provide candid and critical comments that will assist the Climate Change Science Program in making this published report as sound as possible and to ensure that the report meets institutional standards. The peer-review comments, draft manuscript, and response to the peer-review comments are publicly available at: www.climatescience.gov/Library/sap/sap3-1/default.php.

We wish to thank the following individuals for their peer review of this report:
Kerry H. Cook, University of Texas Austin
Carlos R. Mechoso, University of California Los Angeles
Gerald A. Meehl, National Center for Atmospheric Research
Phil Mote, University of Washington Seattle
Brad Udall, Western Water Assessment, Boulder, Colorado
John E. Walsh, International Arctic Research Center

We would also like to thank the following individuals who provided comments during the public comment period:
California Department of Water Resources: Michael Anderson
NOAA Research Council: Derek Parks, Tim Eichler, Michael Winton, Ron Stouffer, and Jiayu Zhou
NOAA Office of Federal Coordination of Meteorology: Samuel P. Williamson
NSF: Marta Cehelsky
The public review comments, draft manuscript, and response to the public comments are publicly available at: www.climatescience.gov/Library/sap/sap3-1/default.php

Intellectual contributions from the following individuals are also acknowledged: John J. Cassano, Elizabeth N. Cassano, Peter Gent, Bala Govindasamy, Xin-Zhong Liang, William Lipscomb, and Thomas J. Phillips.

EDITORIAL TEAM

Technical Editors	Judy Wyrick, Anne Adamson, Oak Ridge National Laboratory
Report Coordinators	Judy Wyrick, Anne Adamson, Shirley Andrews, Oak Ridge National Laboratory
Technical Advisor	David Dokken, CCSPO
Graphic Production	DesignConcept

Recommended Citation for the entire report
CCSP, 2008: *Climate Models: An Assessment of Strengths and Limitations*. A Report by the U.S. Climate Change Science Program and the Subcommittee on Global Change Research [Bader D.C., C. Covey, W.J. Gutowski Jr., I.M. Held, K.E. Kunkel, R.L. Miller, R.T. Tokmakian and M.H. Zhang (Authors)]. Department of Energy, Office of Biological and Environmental Research, Washington, D.C., USA, 124 pp.

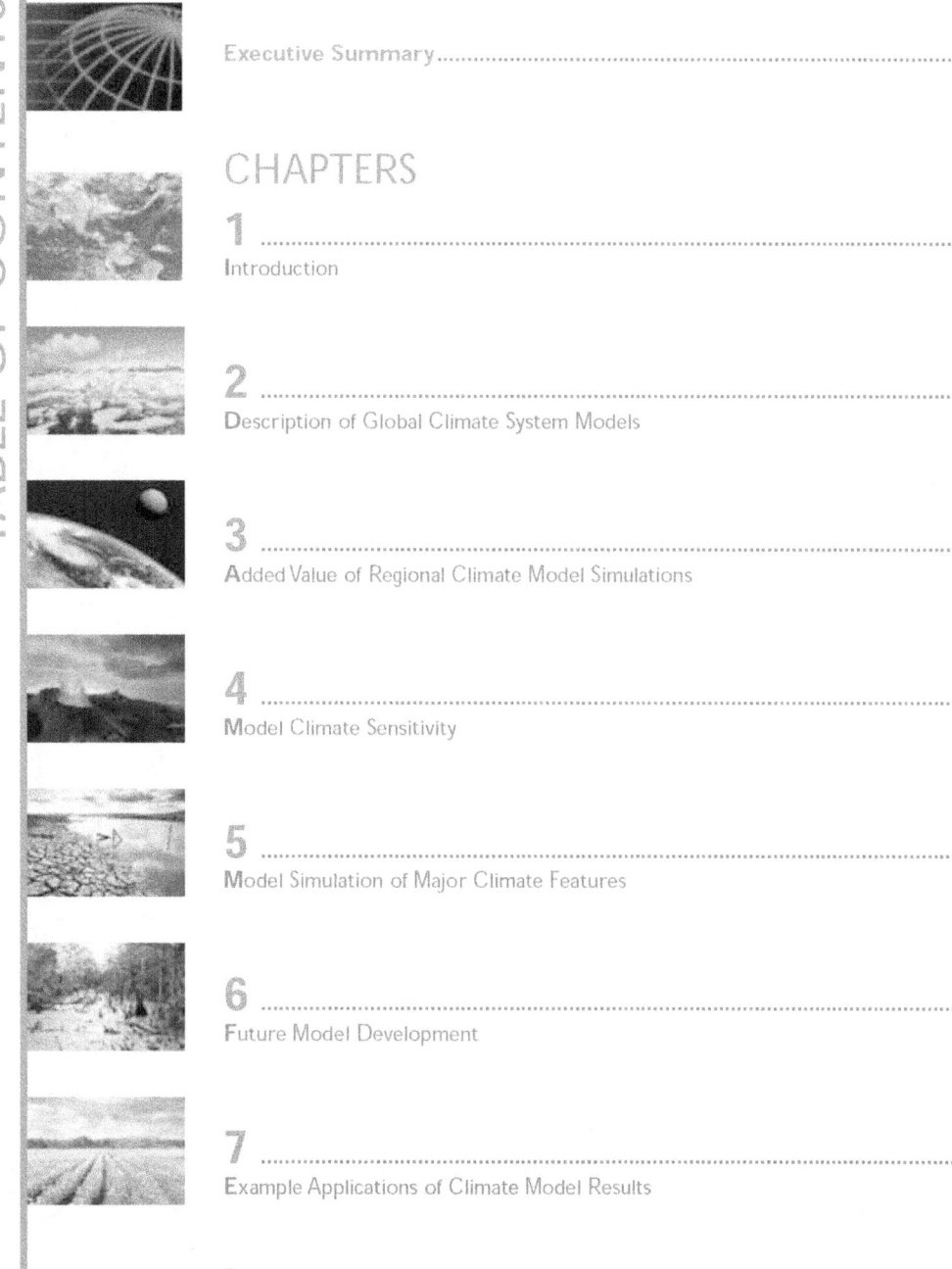

TABLE OF CONTENTS

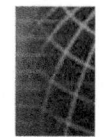

EXECUTIVE SUMMARY

Scientists extensively use mathematical models of Earth's climate, executed on the most powerful computers available, to examine hypotheses about past and present-day climates. Development of climate models is fully consistent with approaches being taken in many other fields of science dealing with very complex systems. These climate simulations provide a framework within which enhanced understanding of climate-relevant processes, along with improved observations, are merged into coherent projections of future climate change. This report describes the models and their ability to simulate current climate.

The science of climate modeling has matured through finer spatial resolution, the inclusion of a greater number of physical processes, and comparison to a rapidly expanding array of observations. These models have important strengths and limitations. They successfully simulate a growing set of processes and phenomena; this set intersects with, but does not fully cover, the set of processes and phenomena of central importance for attribution of past climate changes and the projection of future changes. Following is a concise summary of the information in this report, organized around questions from the "Prospectus," which motivated its preparation, and focusing on these strengths and weaknesses.

What are the major components and processes of the climate system that are included in present state-of-the-science climate models, and how do climate models represent these aspects of the climate system?

Chapter 2 describes the four major components of modern coupled climate models: atmosphere, ocean, land surface, and sea ice. The development of each of these individual components raises important questions as to how key physical processes are represented in models, and some of these questions are discussed in this report. Furthermore, strategies used to couple the components into a climate system model are detailed. Development paths for the three U.S. modeling groups that contributed to the 2007 Intergovernmental Panel on Climate Change (IPCC) Scientific Assessment of Climate Change (IPCC 2007) serve as examples. Experience and expert judgment are essential in constructing and evaluating a climate modeling system, so multiple modeling approaches are

still needed for full scientific evaluation of the state of the science.

The set of most recent climate simulations, referred to as CMIP3 models and utilized heavily in Working Group 1 and 2 reports of the Fourth IPCC Assessment, have received unprecedented scrutiny by hundreds of investigators in various areas of expertise. Although a number of systematic biases are present across the set of models, more generally the simulation strengths and weaknesses, when compared against the current climate, vary substantially from model to model. From many perspectives, an average over the set of models clearly provides climate simulation superior to any individual model, thus justifying the multimodel approach in many recent attribution and climate projection studies.

Climate modeling has been steadily improving over the past several decades, but the pace has been uneven because several important aspects of the climate system present especially severe challenges to the goal of simulation.

How are changes in the Earth's energy balance incorporated into climate models? How sensitive is the Earth's (modeled) climate to changes in the factors that affect the energy balance?

The Earth's radiant energy balance at the top of the atmosphere helps to determine its climate. Chapter 2 contains a brief description of energy-transfer simulation within models, particularly within the atmospheric component. More important, Chapter 4 includes an extensive discussion about radiative forcing of climate change and climate sensitivity. The response of global mean temperature to a doubling of carbon dioxide remains a useful measure of climate sensitivity. The equilibrium response—the response expected after waiting long enough (many hundreds of years) for the system to reequilibrate—is the most commonly quoted measure. Remaining consistent for three decades, the range of equilibrium climate sensitivity obtained from models is roughly consistent with estimates from observations of recent and past climates. The canonical three-fold range of uncertainty, 1.5 to 4.5°C, has evolved very slowly. The lower limit has been nearly unchanged over time, with very few recent models below 2°. Difficulties in simulating Earth's clouds and their response to climate change are the fundamental reasons preventing a reduction in this range in model-generated climate sensitivity.

Other common measures of climate sensitivity measure the climate response on time scales shorter than 100 years. By these measures there is considerably less spread among the models— roughly a factor of two rather than three. The range still is considerable and is not decreasing rapidly, due in part to difficulties in cloud simulation but also to uncertainty in the rate of heat uptake by the oceans. This uncertainty rises in importance when considering the responses on these shorter time scales.

Climate sensitivity in models is subjected to tests using observational constraints. Tests include climate response to volcanic eruptions; aspects of internal climate variability that provide information on the strength of climatic "restoring forces"; the response to the 11-year

cycle in solar irradiance; paleoclimatic information, particularly from the peak of the last Ice Age some 20,000 years ago; aspects of the seasonal cycle; and the magnitude of observed warming over the past century. Because each test is subject to limitations in data and complications from feedbacks in the system, they do not provide definitive tests of models' climate sensitivity in isolation. Studies in which multiple tests of model climate responses are considered simultaneously are essential when analyzing these constraints on sensitivity.

Improvements in our confidence in estimates of climate sensitivity are most likely to arise from new data streams such as the satellite platforms now providing a first look at the three-dimensional global distributions of clouds. New and very computationally intensive climate modeling strategies that explicitly resolve some of the smaller scales of motion influencing cloud cover and cloud radiative properties also promise to improve cloud simulations.

How uncertain are climate model results? In what ways has uncertainty in model-based simulation and prediction changed with increased knowledge about the climate system?

Chapter 1 provides an overview of improvement in models in both completeness and in the ability to simulate observed climate. Climate models are compared to observations of the mean climate in a multitude of ways, and their ability to simulate observed climate changes, particularly those of the past century, have been examined extensively. A discussion of metrics that may be used to evaluate model improvement over time is included at the end of Chapter 2, which cautions that no current model is superior to others in all respects, but rather that different models have differing strengths and weaknesses.

As discussed in Chapter 5, climate models developed in the United States and around the world show many consistent features in their simulations and projections for the future. Accurate simulation of present-day climatology for near-surface temperature and precipitation is necessary for most practical applications of climate modeling. The seasonal cycle and large-scale geographical variations of near-surface temperature are indeed well simulated in recent models, with typical correlations between models and observations of 95% or better.

Climate model simulation of precipitation has improved over time but is still problematic. Correlation between models and observations is 50 to 60% for seasonal means on scales of a few hundred kilometers. Comparing simulated and observed latitude-longitude precipitation maps reveals similarity of magnitudes and patterns in most regions of the globe, with the most striking disagreements occurring in the tropics. In most models, the appearance of the Inter-Tropical Convergence Zone of cloudiness and rainfall in the equatorial Pacific is distorted, and rainfall in the Amazon Basin is substantially underestimated. These errors may prove consequential for a number of model predictions, such as forest uptake of atmospheric CO_2.

Simulation of storms and jet streams in middle latitudes is considered one of the strengths of atmospheric models because the dominant scales involved are reasonably well resolved. As a consequence, there is relatively high confidence in the models' ability to simulate changes in these extratropical storms and jet streams as the climate changes. Deficiencies that still exist may be due partly to insufficient resolution of features such as fronts, to errors in the forcing terms from moist physics, or to inadequacies in simulated interactions between the tropics and midlatitudes or between the stratosphere and the troposphere. These deficiencies are still large enough to impact ocean circulation and some regional climate simulations and projections.

The quality of ocean climate simulations has improved steadily in recent years, owing to better numerical algorithms and more realistic assumptions concerning the mixing occurring on scales smaller than the models' grid. Many of the CMIP3 class of models are able to maintain an overturning circulation in the Atlantic with roughly the observed strength without the artificial correction to air-sea fluxes commonly used in previous generations of models, thus providing a much better foundation for analysis of the circulation's stability. Circulation in the Southern Ocean, thought to be vitally important for oceanic uptake of carbon dioxide from the

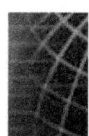

atmosphere, is sensitive to deficiencies in simulated winds and salinities, but a subset of models is producing realistic circulation in the Southern Ocean as well.

Models forced by the observed well-mixed greenhouse gas concentrations, volcanic aerosols, estimates of variations in solar energy incidence, and anthropogenic aerosol concentrations are able to simulate the recorded 20th Century global mean temperature in a plausible way. Solar variations, observed through direct satellite measurements for the last few decades, do not contribute significantly to warming during that period. Solar variations early in the 20th Century are much less certain but are thought to be a potential contributor to warming in that period.

Uncertainties in the climatic effects of man-made aerosols (liquid and solid particles suspended in the atmosphere) constitute a major stumbling block in quantitative attribution studies and in attempts to use the observational record to constrain climate sensitivity. We do not know how much warming due to greenhouse gases has been cancelled by cooling due to aerosols. Uncertainties related to clouds increase the difficulty in simulating the climatic effects of aerosols, since these aerosols are known to interact with clouds and potentially can change cloud radiative properties and cloud cover.

The possibility that natural variability has been a significant contributor to the detailed time evolution seen in the global temperature record is plausible but still difficult to address with models, given the large differences in characteristics of the natural decadal variability between models. While natural variability may very well be relevant to observed variations on the scale of 10 to 30 years, no models show any hint of generating large enough natural, unforced variability on the 100-year time scale to compete with explanations that the observed century-long warming trend has been predominantly forced.

The observed southward displacement of the Southern Hemisphere storm track and jet stream in recent decades is reasonably well simulated in current models, which show that the displacement is due partly to greenhouse gases but also partly to the presence of the stratospheric ozone hole. Circulation changes in the Northern Hemisphere over the past decades have proven more difficult to capture in current models, perhaps because of more complex interactions between the stratosphere and troposphere in the Northern Hemisphere.

Observations of ocean heat uptake are beginning to provide a direct test of aspects of the ocean circulation directly relevant to climate change simulations. Coupled models provide reasonable simulations of observed heat uptake in the oceans but underestimate the observed sea-level rise over the past decades.

Model simulations of trends in extreme weather typically produce global increases in extreme precipitation and severe drought, with decreases in extreme minimum temperatures and frost days, in general agreement with observations.

Simulations from different state-of-the-science models have not fully converged, however, since different groups approach uncertain model aspects in distinctive ways. This absence of convergence is one useful measure of the state of climate simulation; convergence is to be expected once all climate-relevant processes are simulated in a convincing physically based manner. However, measuring the quality of climate models so the metric used is directly relevant to our confidence in the models' projections of future climate has proven difficult. The most appropriate ways to translate simulation strengths and weaknesses into confidence in climate projections remain a subject of active research.

How well do climate models simulate natural variability and how does variability change over time?

Simulation of climate variations also is described in Chapter 5. Simulations of El Niño oscillations, which have improved substantially in recent years, provide a significant success story for climate models. Most current models spontaneously generate El Niño–Southern Oscillation variability, albeit with varying degrees of realism. Oscillation spatial structure and duration are impressive in a model subset but with a tendency toward too short a period. Bias in the Inter-Tropical Convergence Zone (ITCZ) in coupled models is a major factor preventing further improvement in these models. Projections for future El Niño variability and the state of the Pacific Ocean are centrally important for regional climate change projections throughout the tropics and in North America.

Other aspects of the tropical simulations in current models remain inadequate. The Madden-Julian Oscillation, a feature of the tropics in which precipitation is organized by large-scale eastward-propagating features with periods of roughly 30 to 60 days, is a useful test of simulation credibility. Model performance using this measure is still unsatisfactory. The "double ITCZ–cold tongue bias," in which water is excessively cold near the equator and precipitation splits artificially into two zones straddling the equator, remains as a persistent bias in current coupled atmosphere-ocean models. Projections of tropical climate change are affected adversely by these deficiencies in simulations of the organization of tropical convection. Models typically overpredict light precipitation and underpredict heavy precipitation in both the tropics and middle latitudes, creating potential biases when studying extreme events. Tropical cyclones are poorly resolved by the current generation of global models, but recent results with high-resolution atmosphere-only models and dynamical downscaling provide optimism that the simulation of tropical cyclone climatology will advance rapidly in coming years, as will our understanding of observed variations and trends.

The quality of simulations of low-frequency variability on decadal to multidecadal time scales varies regionally and also from model to model. On average, models do reasonably well in the North Pacific and North Atlantic. In other oceanic regions, lack of data contributes to uncertainty in estimating simulation quality at these low frequencies. A dominant mode of low-frequency variability in the atmosphere, known as northern and southern annular modes, is very well captured in current models. These modes involve north-south displacements of the extratropical storm track and have dominated observed atmospheric circulation trends in recent decades. Because of their ability to simulate annular modes, global climate models do fairly well with interannual variability in polar regions of both hemispheres. They are less successful with daily polar-weather variability, although finer-scale regional simulations do show promise for improved global-model simulations as their resolution increases.

How well do climate models simulate regional climate variability and change?

Chapter 3 describes techniques to downscale coarse-resolution global climate model output to higher resolution for regional applications. These downscaling methodologies fall primarily into two categories. In the first, a higher-resolution, limited-area numerical meteorological model is driven by global climate model output at its lateral boundaries. These dynamical downscaling strategies are beneficial when supplied with appropriate sea-surface and atmospheric boundary conditions, but their value is limited by uncertainties in information supplied by global models. Given the value of multimodel ensembles for larger-scale climate prediction, coordinated downscaling clearly must be performed with a representative set of global model simulations as input, rather than focusing on results from one or two models. Relatively few such multimodel dynamical downscaling studies have been performed to date.

In the second category, empirical relationships between large- and small-scale observations are developed, then applied to global climate model output to provide regional detail. Statistical techniques to produce appropriate small-scale structures from climate simulations are referred to as "statistical downscaling." They can be as effective as high-resolution numerical simulations in providing climate change information to regions unresolved by most current global models. Because of the computational efficiency of these techniques, they can much more easily utilize a full suite of multimodel ensembles. The statistical methods, however, are completely dependent on the accuracy of regional circulation patterns produced by global models. Dynamical models, through higher resolution or better representation of important physical processes, often can improve the physical realism of simulated regional circulation. Thus,

the strengths and weaknesses of dynamical modeling and statistical methods often are complementary.

Regional trends in extreme events are not always captured by current models, but it is difficult to assess the significance of these discrepancies and to distinguish between model deficiencies and natural variability.

The use of climate model results to assess economic, social, and environmental impacts is becoming more sophisticated, albeit slowly. Simple methods requiring only mean changes in temperature and precipitation to estimate impacts remain popular, but an increasing number of studies are using more detailed information such as the entire distribution of daily or monthly values and extreme outcomes. The mismatch between models' spatial resolution vs the scale of impact-relevant climate features and of impact models remains an impediment for certain applications. Chapter 7 provides several examples of applications using climate model results and downscaling techniques.

What are the tradeoffs to be made in further climate model development (e.g., between increasing spatial/temporal resolution and representing additional physical/biological processes)?

Chapter 6 is devoted to trends in climate model development. With increasing computer power and observational understanding, future models will include both higher resolution and more processes.

Resolution increases most certainly will lead to improved representations of atmospheric and oceanic general circulations. Ocean components of current climate models do not directly simulate the oceans' very energetic motions referred to as "mesoscale eddies." Simulation of these small-scale flow patterns requires horizontal grid sizes of 10 km or smaller. Current oceanic components of climate models are effectively laminar rather than turbulent, and the effects of these eddies must be approximated by imperfect theories. As computer power increases, new models that resolve these eddies will be incorporated into climate models to explore their impact on decadal variability as well as heat and carbon uptake. Similarly, atmospheric general circulation models will evolve to "cloud-resolving models" (CRMs) with spatial resolutions of less than a few kilometers. The hope is that CRMs will provide better results through explicit simulation of many cloud properties now poorly represented on subgrid scales of current atmospheric models. CRMs are not new frameworks but rather are based on models designed for mesoscale storm and cumulus convection simulations.

Models of glacial ice are in their infancy. Glacial models directly coupled to atmosphere-ocean models typically account for only direct melting and accumulation at the surface of ice sheets and not the dynamic discharge due to glacial flow. More-detailed current models typically generate discharges that change only over centuries and millennia. Recent evidence for rapid variations in this glacial outflow indicates that more-realistic glacial models are needed to estimate the evolution of future sea level.

Inclusion of carbon-cycle processes and other biogeochemical cycles is required to transform physical climate models into full Earth system models that incorporate feedbacks influencing greenhouse gas and aerosol concentrations in the atmosphere. Land models that predict vegetation patterns are being developed actively, but the demands of these models on the quality of simulated precipitation patterns ensures that their evolution will be gradual and tied to improvements in the simulation of regional climate. Uncertainties about carbon-feedback processes in the ocean as well as on land, however, must be reduced for more reliable future estimates of climate change.

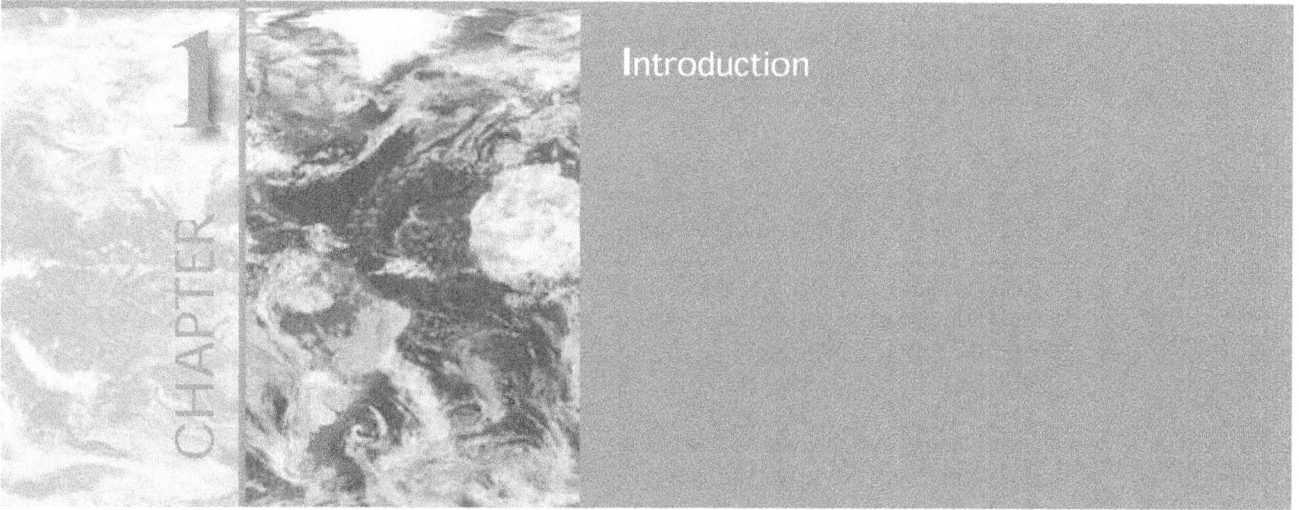

Introduction

The use of computers to simulate complex systems has grown in the past few decades to play a central role in many areas of science. Climate modeling is one of the best examples of this trend and one of the great success stories of scientific simulation. Building a laboratory analog of the Earth's climate system with all its complexity is impossible. Instead, the successes of climate modeling allow us to address many questions about climate by experimenting with simulations—that is, with mathematical models of the climate system. Despite the success of the climate modeling enterprise, the complexity of our Earth imposes important limitations on existing climate models. This report aims to help the reader understand the valid uses, as well as the limitations, of current climate models.

Climate modeling and forecasting grew from the desire to predict weather. The distinction between climate and weather is not precise. Operational weather forecasting has focused historically on time scales of a few days but more recently has been extended to months and seasons in attempts to predict the evolution of El Niño episodes. The goal of climate modeling can be thought of as the extension of forecasting to longer and longer time periods. The focus is not on individual weather events, which are unpredictable on long time scales, but on the statistics of these events and on the slow evolution of oceans and ice sheets. Whether the forecasting of individual El Niño episodes is considered weather or climate is a matter of convention. For the purpose of this report, we will consider El Niño forecasting as weather and will not address it directly. On the climate side we are concerned, for example, with the ability of models to simulate the statistical characteristics of El Niño variability or extratropical storms or Atlantic hurricanes, with an eye toward assessing the ability of models to predict how variability might change as the climate evolves in coming decades and centuries.

An important constraint on climate models not imposed on weather-forecast models is the requirement that the global system precisely and accurately maintain the global energy balance over very long periods of time. The Earth's energy balance (or "budget") is defined as the difference between absorbed solar energy and emitted infrared radiation to space. It is affected by many factors, including the accumulation of greenhouse gases, such as carbon dioxide, in the atmosphere. The decades-to-century changes in the Earth's energy budget, manifested as climate changes, are just a few percent of the average values of that budget's largest terms. Many decisions about model construction described in

Chapter 2 are based on the need to properly and accurately simulate the long-term energy balance.

This report will focus primarily on comprehensive physical climate models used for the most recent international Coupled Model Intercomparison Project (CMIP) coordinated experiments (Meehl et al. 2006) sponsored by the World Climate Research Programme (WCRP). These coupled atmosphere-ocean general circulation models (AOGCMs) incorporate detailed representations of the atmosphere, land surface, oceans, and sea ice. Where practical, we will emphasize and highlight results from the three U.S. modeling projects that participated in the CMIP experiments. Additionally, this report examines the use of regional climate models (RCMs) for obtaining higher-resolution details from AOGCM simulations over smaller regions. Still, other types of climate models are being developed and applied to climate simulation. The more-complete Earth system models, which build carbon-cycle and ecosystem processes on top of AOGCMs, are used primarily for studies of future climate change and paleoclimatology, neither of which is directly relevant to this report. Another class of models not discussed here but used extensively, particularly when computer resources are limited, is Earth system models of intermediate complexity (EMICs). Although these models have many more assumptions and simplifications than are found in CMIP models (Claussen et al. 2002), they are particularly useful in exploring a wide range of mechanisms and obtaining broad estimates of future climate change projections that can be further refined with AOGCM experiments.

1.1 BRIEF HISTORY OF CLIMATE MODEL DEVELOPMENT

As numerical weather prediction was developing in the 1950s as one of the first computer applications, the possibility of also using numerical simulation to study climate became evident almost immediately. The feasibility of generating stable integrations of atmospheric equations for arbitrarily long time periods was demonstrated by Norman Phillips in 1956. About that time, Joseph Smagorinsky started a program in climate modeling that ultimately be-

came one of the most vigorous and longest-lived GCM development programs at the National Oceanic and Atmospheric Administration's Geophysical Fluid Dynamics Laboratory (GFDL) at Princeton University. The University of California at Los Angeles began producing atmospheric general circulation models (AGCMs) beginning in 1961 under the leadership of Yale Mintz and Akio Arakawa. This program influenced others in the 1960s and 1970s, leading to modeling programs found today at National Aeronautics and Space Administration (NASA) laboratories and several universities. At Lawrence Livermore National Laboratory, Cecil E. Leith developed an early AGCM in 1964. The U.S. National Center for Atmospheric Research (NCAR) initiated AGCM development in 1964 under Akira Kasahara and Warren Washington. Leith moved to NCAR in the late 1960s and, in the early 1980s, oversaw construction of the Community Climate Model, a predecessor to the present Community Climate System Model (CCSM).

Early weather models focused on fluid dynamics rather than on radiative transfer and the atmosphere's energy budget, which are centrally important for climate simulations. Additions to the original AGCMs used for weather analysis and prediction were needed to make climate simulations possible. Furthermore, because climate simulation focuses on time scales longer than a season, oceans and sea ice must be included in the modeling system in addition to the more rapidly evolving atmosphere. Thus, ocean and ice models have been coupled with atmospheric models. The first ocean GCMs were developed at GFDL by Bryan and Cox in the 1960s and then coupled with the atmosphere by Manabe and Bryan in the 1970s. Paralleling events in the United States, the 1960s and 1970s also were a period of climate- and weather-model development throughout the world, with major centers emerging in Europe and Asia. Representatives of these groups gathered in Stockholm in August 1974, under the sponsorship of the Global Atmospheric Research Programme to produce a seminal treatise on climate modeling (GARP 1975). This meeting established collaborations that still promote international cooperation today.

The use of climate models in research on carbon dioxide and climate began in the early 1970s. The important study, "Inadvertent Climate Modification" (SMIC 1971), endorsed the use of GCM-based climate models to study the possibility of anthropogenic climate change. With continued improvements in both climate observations and computer power, modeling groups furthered their models through steady but incremental improvements. By the late 1980s, several national and international organizations formed to assess and expand scientific research related to global climate change. These developments spurred interest in accelerating the development of improved climate models. The primary focus of Working Group 1 of the United Nations Intergovernmental Panel on Climate Change (IPCC), which began in 1988, was the scientific inquiry into physical processes governing climate change. IPCC's first Scientific Assessment (IPCC 1990) stated, "Improved prediction of climate change depends on the development of climate models, which is the objective of the climate modeling programme of the World Climate Research Programme." The United States Global Change Research Program (USGCRP), established in 1989, designated climate modeling and prediction as one of the four high-priority integrating themes of the program (Our Changing Planet 1991). The combination of steadily increasing computer power and research spurred by WCRP and USGCRP has led to a steady improvement in the completeness, accuracy, and resolution of AOGCMS for climate simulation and prediction. An often-used illustration from the Third IPCC Working Group 1 *Scientific Assessment of Climate Change* in 2001 depicts this evolution (see Fig. 1.1). Even more comprehensive climate models produced a series of coordinated numerical simulations for the third international Climate Model Intercomparison Project (CMIP3), which were used extensively in research cited in the recent Fourth IPCC Assessment (IPCC 2007). Contributions came from three groups in the United States (GFDL, NCAR, and the NASA Goddard Institute for Space Studies) and others in the United Kingdom, Germany, France, Japan, Australia, Canada, Russia, China, Korea, and Norway.

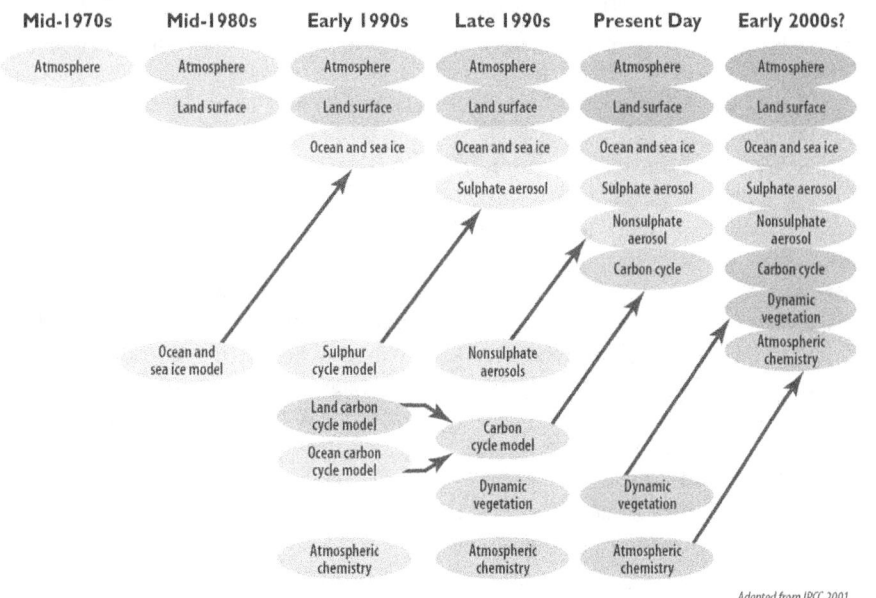

Development of Climate Models: Past, Present, and Future

Adapted from IPCC 2001

Figure 1.1. Historical Development of Climate Models.

[Figure source: *Climate Change 2001: The Scientific Basis, Contribution of Working Group 1 to the Assessment Report of the Intergovernmental Panel on Climate Change*, p. 48. Used with permission from IPCC.]

1.2 CLIMATE MODEL CONSTRUCTION

Comprehensive climate models are constructed using expert judgments to satisfy many constraints and requirements. Overarching considerations are the accurate simulation of the most important climate features and the scientific understanding of the processes that control these features. Typically, the basic requirement is that models should simulate features important to humans, particularly surface variables such as temperature, precipitation, windiness, and storminess. This is a less-straightforward requirement than it seems because a physically based climate model also must simulate all complex interactions in the coupled atmosphere–ocean–land surface–ice system manifested as climate variables of interest. For example, jet streams at altitudes of 10 km above the surface must be simulated accurately if models are to generate midlatitude weather with realistic characteristics. Midlatitude highs and lows shown on surface weather maps are intimately associated with these high-altitude wind patterns. As another example, the basic temperature decrease from the equator to the poles cannot be simulated without taking into account the poleward transport of heat in the oceans, some of this heat being carried by currents 2 or 3 km deep into the ocean interior. Thus, comprehensive models should produce correctly not just the means of variables of interest but also the extremes and other measures of natural variability. Finally, our models should be capable of simulating changes in statistics caused by relatively small changes in the Earth's energy budget that result from natural and human actions.

Climate processes operate on time scales ranging from several hours to millennia and on spatial scales ranging from a few centimeters to thousands of kilometers. Principles of scale analysis, fluid dynamical filtering, and numerical analysis are used for intelligent compromises and approximations to make possible the formulation of mathematical representations of processes and their interactions. These mathematical models are then translated into computer codes executed on some of the most powerful computers in the world. Available computer power helps determine the types of approximations required. As a general rule, growth of computational resources allows modelers to formulate algorithms less dependent on approximations known to have limitations, thereby producing simulations more solidly founded on established physical principles. These approximations are most often found in "closure" or "parameterization" schemes that take into account unresolved motions and processes and are always required because climate simulations must be designed so they can be completed and analyzed by scientists in a timely manner, even if run on the most powerful computers.

Climate models have shown steady improvement over time as computer power has increased, our understanding of physical processes of climatic relevance has grown, datasets useful for model evaluation have been developed, and our computational algorithms have improved. Figure 1.2 shows one attempt at quantifying this change. It compares a particular metric of climate model performance among the CMIP1 (1995), CMIP2 (1997), and CMIP3 (2004) ensembles of AOGCMs. This particular metric assesses model performance in simulating the mean climate of the late 20th Century as measured by a basket of indicators focusing on aspects of atmospheric climate for which observational counterparts are deemed adequate. Model ranking according to individual members of this basket of indicators varies greatly, so this aggregate ranking depends on how different indicators are weighted in relative importance. Nevertheless, the conclusion that models have improved over time is not dependent on the relative weighting factors, as nearly all models have improved in most respects. The construction of metrics for evaluating climate models is itself a subject of intensive research and will be covered in more detail in Chapter 2.

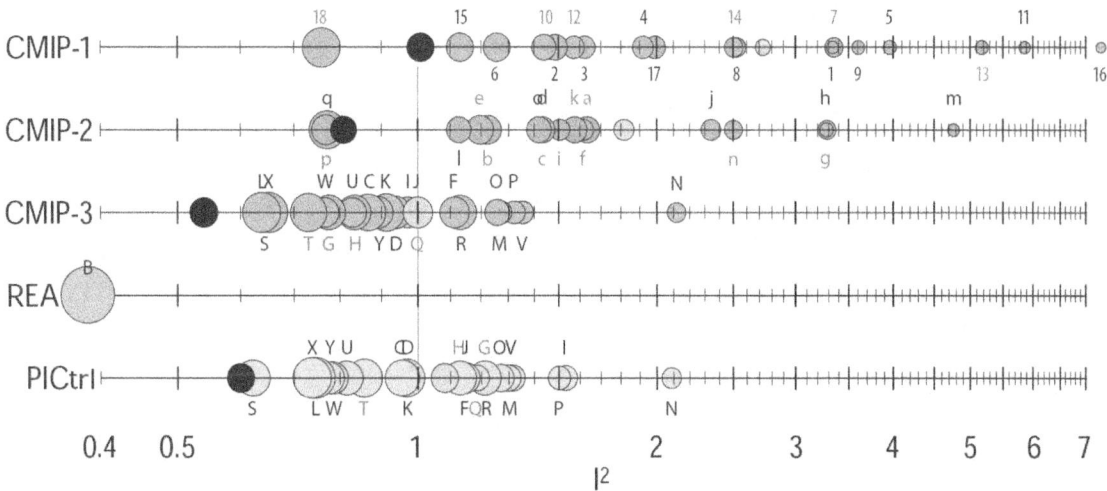

Figure 1.2. Performance Index *I²* for Individual Models (circles) and Model Generations (rows).

Best performing models have low *I²* values and are located toward the left. Circle sizes indicate the length of the 95% confidence intervals. Letters and numbers identify individual models; flux corrected models are labeled in red. Grey circles show the average *I²* of all models within one model group. Black circles indicate the *I²* of the multimodel mean taken over one model group. The green circle (REA) corresponds to the *I²* of the NCEP/NCAR Reanalysis (Kalnay et al. 1996), conducted by the National Weather Service's National Centers for Environmental Prediction and the National Center for Atmospheric Research. Last row (PICTRL) shows *I²* for the preindustrial control experiment of the CMIP3 project. [Adapted from Fig. 1 in T. Reichler and J. Kim 2008: How well do coupled models simulate today's climate? *Bulletin American Meteorological Society,* **89**(3), doi:10.1175/BAMS-89-3-303. Reproduced by permission of the American Meteorological Society.]

Also shown in Fig. 1.2 is the same metric evaluated from climate simulation results obtained by averaging over all AOGCMs in the CMIP1, CMIP2, and CMIP3 archives. The CMIP3 "ensemble-mean" model performs better than any individual model by this metric and by many others. This kind of result has convinced the community of the value of a multimodel approach to climate change projection. Our understanding of climate is still insufficient to justify proclaiming any one model "best" or even showing metrics of model performance that imply skill in predicting the future. More appropriate in any assessments focusing on adaptation or mitigation strategies is to take into account, in a pertinently informed manner, the products of distinct models built using different expert judgments at centers around the world.

1.3 SUMMARY OF SAP 3.1 CHAPTERS

The remaining sections of this report describe climate model development, evaluation, and applications in more detail. Chapter 2 describes the development and construction of models and how they are employed for climate research. Chapter 3 discusses regional climate models and their use in "downscaling" global model results to specific geographic regions, particularly North America. The concept of climate sensitivity—the response of a surface temperature to a specified change in the energy budget at the top of the model's atmosphere—is described in Chapter 4. A survey of how well important climate features are simulated by modern models is found in Chapter 5, while Chapter 6 depicts near-term development priorities for future model development. Finally, Chapter 7 illustrates a few examples of how climate model simulations are used for practical applications. A detailed Reference section follows Chapter 7.

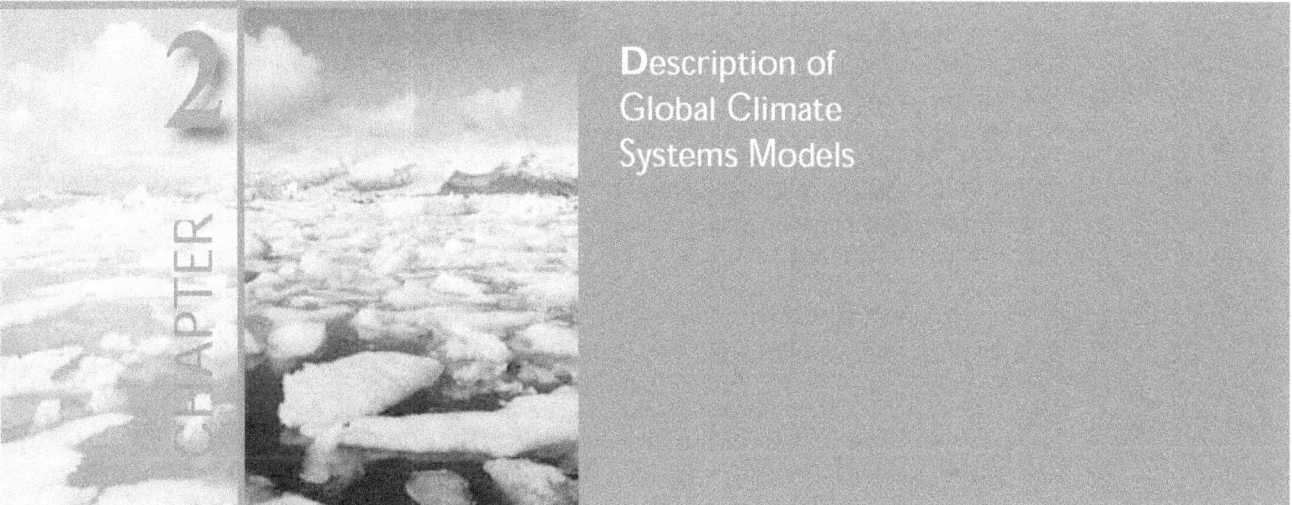

Description of Global Climate Systems Models

CHAPTER 2

Modern climate models are composed of a system of interacting model components, each of which simulates a different part of the climate system. The individual parts often can be run independently for certain applications. Nearly all the Coupled Model Intercomparison Project 3 (CMIP3) class of models include four primary components: atmosphere, land surface, ocean, and sea ice. The atmospheric and ocean components are known as "general circulation models" or GCMs because they explicitly simulate the large-scale global circulation of the atmosphere and ocean. Climate models sometimes are referred to as coupled atmosphere-ocean GCMs. This name may be misleading because coupled GCMs can be employed to simulate aspects of weather and ocean dynamics without being able to maintain a realistic climate projection over centuries of simulated time, as required of a climate model used for studying anthropogenic climate change. What follows in this chapter is a description of a modern climate model's major components and how they are coupled and tested for climate simulation.

2.1 ATMOSPHERIC GENERAL CIRCULATION MODELS

Atmospheric general circulation models (AGCMs) are computer programs that evolve the atmosphere's three-dimensional state forward in time. This atmospheric state is described by such variables as temperature, pressure, humidity, winds, and water and ice condensate in clouds. These variables are defined on a spatial grid, with grid spacing determined in large part by available computational resources. Some processes governing this atmospheric state's evolution are relatively well resolved by model grids and some are not. The latter are incorporated into models through approximations often referred to as parameterizations. Processes that transport heat, water, and momentum horizontally are relatively well resolved by the grid in current atmospheric models, but processes that redistribute these quantities vertically have a significant part that is controlled by subgrid-scale parameterizations.

The model's grid-scale evolution is determined by equations describing the thermodynamics and fluid dynamics of an ideal gas. The atmosphere is a thin spherical shell of air that envelops the Earth. For climate simulation, emphasis is placed on the atmosphere's lowest 20 to 30 km (i.e., the troposphere and the lower stratosphere). This layer contains over 95% of the atmosphere's mass and virtually all of its water vapor, and it produces nearly all weather although current research suggests possible interactions between this layer and higher atmospheric levels (e.g., Pawson et al. 2000). Because of the disparity between scales of horizontal and vertical motions governing global and regional climate, the two motions are treated differently by model algorithms. The resulting set of equations is often referred to as the primitive equations (Haltiner and Williams 1980).

Although nearly all AGCMs use this same set of primitive dynamical equations, they use different numerical algorithms to solve them. In all cases, the atmosphere is divided into discrete vertical layers, which are then overlaid with a two-dimensional horizontal grid, producing a three-dimensional mesh of grid elements. The equations are solved as a function of time on this mesh. The portion of the model code governing the fluid dynamics explicitly simulated on this mesh often is referred to as the model's "dynamical core." Even with the same numerical approach, AGCMs differ in spatial resolutions and configuration of model grids. Some models use a "spectral" representation of winds and temperatures, in which these fields are written as linear combinations of predefined patterns on the sphere (spherical harmonics) and are then mapped to a grid when local values are required. Some models have few layers above the tropopause (the moving boundary between the troposphere and stratosphere (e.g., GFDL 2004)), while others have as many layers above the troposphere as in it (e.g., Schmidt et al. 2006).

All AGCMs use a coordinate system in which the Earth's surface is a coordinate surface, simplifying exchanges of heat, moisture, trace substances, and momentum between the Earth's surface and the atmosphere. Numerical algorithms of AGCMs should precisely conserve the atmosphere's mass and energy. Typical AGCMs have spatial resolution of 200 km in the horizontal and 20 levels in the volume below the altitude of 15 km. Because numerical errors often depend on flow patterns, there are no simple ways to assess the accuracy of numerical discretizations in AGCMs. Models use idealized cases testing the model's long-term stability and efficiency (e.g., Held and Suarez 1994), as well as tests focusing on accuracy using short integrations (e.g., Polvani, Scott, and Thomas 2005).

All AGCMs must incorporate the effects of radiant-energy transfer. The radiative-transfer code computes the absorption and emission of electromagnetic waves by air molecules and atmospheric particles. Atmospheric gases absorb and emit radiation in "spectral lines" centered at discrete wavelengths, but the computational costs are too high in a climate model to perform this calculation for each individual spectral line. AGCMs use approximations, which differ among models, to group bands of wavelengths together in a more efficient calculation. Most models have separate radiation codes to treat solar (visible) radiation and the much-longer-wavelength terrestrial (infrared) radiation. Radiation calculation includes the effects of water vapor, carbon dioxide, ozone, and clouds. Models used in climate change experiments also include aerosols and additional trace gases such as methane, nitrous oxide, and the cloroflourocarbons. Validation of AGCM radiation codes often is done offline (separate from other AGCM components) by comparison with line-by-line model calculations that, in turn, are compared against laboratory and field observations (e.g., Ellingson and Fouquart 1991; Clough, Iacono, and Moncet 1992; Collins et al. 2006b).

All GCMs use subgrid-scale parameterizations to simulate processes that are too small or operate on time scales too fast to be resolved on the model grid. The most important parameterizations are those involving cirrus and stratus cloud formation and dissipation, cumulus convection (thunderstorms and fair-weather cumulus clouds), and turbulence and subgrid-scale mixing. For cloud calculations, most AGCMs treat ice and liquid water as atmospheric state variables. Some models also separate cloud particles into ice crystals, snow, graupel (snow pellets), cloud water, and rainwater. Empirical relationships are used to calculate conversions among different particle types. Representing these processes on the scale of model grids is particularly difficult and involves calculation of fractional cloud cover within a grid box, which greatly affects radiative transfer and model sensitivity. Models either predict cloud amounts from the instantaneous thermodynamical and hydrological state of a grid box or they treat cloud fraction as a time-evolving model variable. In higher-resolution models, one can attempt to explicitly simulate the size distribution of cloud particles and the "habit" or nonspherical shape of ice particles, but no current global AGCMs attempt this.

Cumulus convective transports, which are important in the atmosphere but cannot be explicitly resolved at GCM scale, are calculated using convective parameterization algorithms. Most current models use a cumulus mass flux scheme patterned after that proposed by Arakawa and Schubert (1974), in which convection's upward motion occurs in very narrow plumes that take up a negligible fraction of a grid box's area. Schemes differ in techniques used to determine the amount of mass flowing through these plumes and the manner in which air is entrained and detrained by the rising plume. Most models do not calculate separately the area and vertical velocity of convection but try to predict only the product of mass and area, or convective mass flux. Prediction of convective velocities, however, is needed for new models of interactions between aerosols and clouds. Most current schemes do not account for differences between organized mesoscale convective systems and simple plumes. The turbulent mixing rate of updrafts and downdrafts with environments and the phase changes of water vapor within convective systems are treated with a mix of empiricism and constraints based on the moist thermodynamics of rising air parcels. Some models also include a separate parameterization of shallow, nonprecipitating convection (fairweather cumulus clouds). In short, clouds generated by cumulus convection in climate models should be thought of as based in large part on empirical relationships.

All AGCMs parameterize the turbulent transport of momentum, moisture, and energy in the atmospheric boundary layer near the surface. A long-standing theoretical framework, Monin-Obukhov similarity theory, is used to calculate the vertical distribution of turbulent fluxes and state variables in a thin (typically less than 10 m) layer of air adjacent to the surface. Above the surface layer, turbulent fluxes are calculated based on closure assumptions that provide a complete set of equations for subgrid-scale variations. Closure assumptions differ among AGCMs; some models use high-order closures

in which the fluxes or second-order moments are calculated prognostically (with memory in these higher-order moments from one time step to the next). Turbulent fluxes near the surface depend on surface conditions such as roughness, soil moisture, and vegetation. In addition, all models use diffusion schemes or dissipative numerical algorithms to simulate kinetic energy dissipation from turbulence far from the surface and to damp small-scale unresolved structures produced from resolved scales by turbulent atmospheric flow.

The realization that a significant fraction of momentum transfer between atmosphere and surface takes place through nonturbulent pressure forces on small-scale "hills" has resulted in a substantial effort to understand and model this transfer (e.g., McFarlane 1987; Kim and Lee 2003). This process is often referred to as gravity wave drag because it is intimately related to atmospheric wave generation. The variety of gravity wave drag parameterizations is a significant source of differences in mean wind fields generated by AGCMs. Accounting for both surface-generated and convectively generated gravity waves are difficult aspects of modeling the stratosphere and mesosphere (≥ 20 km altitude), since winds in those regions are affected strongly by transfer of momentum and energy from these unresolved waves.

Extensive field programs have been designed to evaluate parameterizations in GCMs, ranging from tests of gravity wave drag schemes [Mesoscale Alpine Program (called MAP), e.g., Bougeault et al. 2001] to tests of radiative transfer and cloud parameterizations [Atmospheric Radiation Measurement Program (called ARM), Ackerman and Stokes 2003]. Running an AGCM coupled to a land model as a numerical weather prediction model for a few days—starting with best estimates of the atmosphere and land's instantaneous state at any given time—is a valuable test of the entire package of atmospheric parameterizations and dynamical core (e.g., Xie et al. 2004). Atmosphere-land models also are routinely tested by running them with boundary conditions taken from observed sea-surface temperatures and sea-ice distributions (Gates 1992) and examining the resulting climate.

2.2 OCEAN GENERAL CIRCULATION MODELS

Ocean general circulation models (OGCMs) solve the primitive equations for global incompressible fluid flow analogous to the ideal-gas primitive equations solved by atmospheric GCMs. In climate models, OGCMs are coupled to the atmosphere and ice models through the exchange of heat, salinity, and momentum at the boundary among components. Like the atmosphere, the ocean's horizontal dimensions are much larger than its vertical dimension, resulting in separation between processes that control horizontal and vertical fluxes. With continents, enclosed basins, narrow straits, and submarine basins and ridges, the ocean has a more complex three-dimensional boundary than does the atmosphere.. Furthermore, the thermodynamics of sea water is very different from that of air, so an empirical equation of state must be used in place of the ideal gas law.

An important distinction among ocean models is the choice of vertical discretization. Many models use vertical levels that are fixed distances below the surface (Z-level models) based on the early efforts of Bryan and Cox (1967) and Bryan (1969a, b). The General Fluid Dynamics Laboratory (GFDL) and Community Climate System Model (CCSM) ocean components fall into this category (Griffies et al. 2005; Maltrud et al. 1998). Two Goddard Institute for Space Studies (GISS) models (R and AOM) use a variant of this approach in which mass rather than height is used as the vertical coordinate (Russell, Miller, and Rind 1995; Russell et al. 2000). A more fundamental alternative uses density as a vertical coordinate. Motivating this choice is the desire to control as precisely as possible the exchange of heat between layers of differing density, which is very small in much of the ocean yet centrally important for simulation of climate. The GISS EH model utilizes a hybrid scheme that transitions from a Z-coordinate near the surface to density layers in the ocean interior (Sun and Bleck 2001; Bleck 2002; Sun and Hansen 2003).

Horizontal grids used by most ocean models in the CMIP3 archive are comparable to or somewhat finer than grids in the atmospheric models

to which they are coupled, typically on the order of 100 km (~ 1° spacing in latitude and longitude) for most of Earth. In many OGCMs the north-south resolution is enhanced within 5° latitude of the equator to improve the ability to simulate important equatorial processes. OGCM grids usually are designed to avoid coordinate singularities caused by the convergence of meridians at the poles. For example, the CCSM OGCM grid is rotated to place its North Pole over a continent, while the GFDL models use a grid with three poles, all of which are placed over land (Murray 1996). Such a grid results in having all ocean grid points at numerically viable locations.

Processes that control ocean mixing near the surface are complex and take place on small scales (order of centimeters). To parameterize turbulent mixing near the surface, the current generation of OGCMs uses several different approaches (Large, McWilliams, and Doney 1994) similar to those developed for atmospheric near-surface turbulence. Within the ocean's stratified, adiabatic interior, vertical mixing takes place on scales from meters to kilometers (Fig. 2.1); the smaller scales also must be parameterized in ocean components. Ocean mixing contributes to its heat uptake and stratification, which in turn affects circulation patterns over time scales of decades and longer. Experts generally feel (e.g., Schopf et al. 2003) that subgrid-scale mixing parameterizations in OGCMs contribute significantly to uncertainty in estimates of the ocean's contribution to climate change.

Very energetic eddy motions occur in the ocean on the scale of a few tens of kilometers. These so-called mesoscale eddies are not present in the ocean simulations of CMIP3 climate models. Ocean models used for climate simulation cannot afford the computational cost of explicitly resolving ocean mesoscale eddies. Instead, they must parameterize mixing by the eddies. Treatment of these mesoscale eddy effects is an important factor distinguishing one ocean model from another. Most real ocean mixing is along rather than across surfaces of constant density. Development of parameterizations that account for this essential feature of mesoscale eddy mixing (Gent and McWilliams 1990; Griffies 1998)

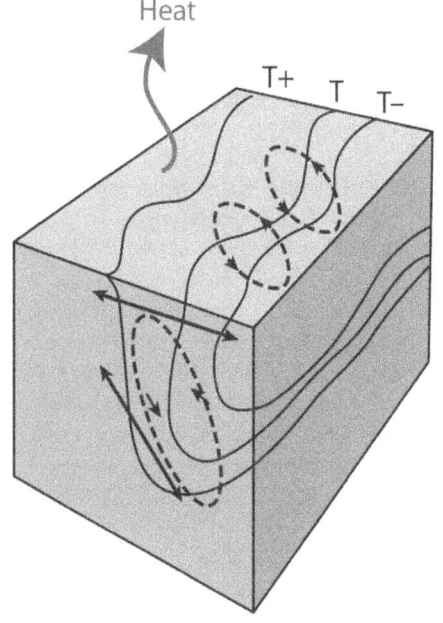

Figure 2.1. Schematic Showing Interaction of a Well-Mixed Surface Layer with Stratified Interior in a Region with a Strong Temperature Gradient.

Mixing (dashed lines) is occurring both across temperature (T) gradients and along the temperature gradient with increasing depth. This process is poorly observed and not well understood. It must be parameterized in large-scale models. [Adapted from Fig. 1, p. 18, in *Coupling Process and Model Studies of Ocean Mixing to Improve Climate Models—A Pilot Climate Process Modeling and Science Team*, a U.S. CLIVAR white paper by Schopf et al. (2003). Figure originated by John Marshall, Massachusetts Institute of Technology.]

is a major advance in recent ocean and climate modeling. Inclusion of higher-resolution, mesoscale eddy–resolving ocean models in future climate models would reduce uncertainties associated with these parameterizations.

Other mixing processes that may be important in the ocean include tidal mixing and turbulence generated by interactions with the ocean's bottom, both of which are included in some models. Lee, Rosati, and Spellman (2006) describe some effects of tidal mixing in a climate model. Some OGCMs also explicitly treat the bottom boundary and sill overflows (Beckman and Dosher 1997; Roberts and Wood 1997; Griffies et al. 2005). Furthermore, sunlight penetration into the ocean is controlled by chlorophyll distributions (e.g., Paulson and Simpson 1977; Morel and Antoine 1994; Ohlmann 2003), and the depth of penetration can affect surface temperatures. All U.S. CMIP3 models include some treatment of this effect, but they prescribe rather than attempt to simulate the upper ocean biology controlling water opacity. Finally, the inclusion of fresh water input by rivers is essential to close the global hydrological cycle; it affects ocean mixing locally and is handled by models in a variety of ways.

The relatively crude resolution of OGCMs used in climate models results in isolation of the smaller seas from large ocean basins. This re-

quires models to perform ad hoc exchanges of water between the isolated seas and the ocean to simulate what in nature involves a channel or strait. (The Strait of Gibraltar is an excellent example.) Various modeling groups have chosen different methods to handle water mixing between smaller seas and larger ocean basins.

OGCM components of climate models are often evaluated in isolation—analogous to the evaluation of AGCMs with prescribed ocean and sea-ice boundary conditions—in addition to being evaluated as components of fully coupled ocean-atmosphere GCMs. (Results of full AOGCM evaluation are discussed in Chapter 5.) Evaluation of ocean models in isolation requires input of boundary conditions at the air-sea interface. To compare simulations with observed data, boundary conditions or surface forcing are from the same period as the data. These surface fluxes also have uncertainties and, as a result, the evaluation of OGCMs with specified sea-surface boundary conditions must take these uncertainties into account.

2.3 LAND-SURFACE MODELS

Interaction of Earth's surface with its atmosphere is an integral aspect of the climate system. Exchanges (fluxes) of mass and energy, water vapor, and momentum occur at the interface. Feedbacks between atmosphere and sur-

face affecting these fluxes have important effects on the climate system (Seneviratne et al. 2006). Modeling the processes taking place over land is particularly challenging because the land surface is very heterogeneous and biological mechanisms in plants are important. Climate model simulations are very sensitive to the choice of land models (Irannejad, Henderson-Sellers, and Sharmeen 2003).

In the earliest global climate models, land-surface modeling occurred in large measure to provide a lower boundary to the atmosphere that was consistent with energy, momentum, and moisture balances (e.g., Manabe 1969). The land surface was represented by a balance among incoming and outgoing energy fluxes and a "bucket" that received precipitation from the atmosphere and evaporated moisture into the atmosphere, with a portion of the bucket's water draining away from the model as a type of runoff. The bucket's depth equaled soil field capacity. Little attention was paid to the detailed set of biological, chemical, and physical processes linked together in the climate system's terrestrial portion. From this simple starting point, land surface modeling for climate simulation has increased markedly in sophistication, with increasing realism and inclusiveness of terrestrial surface and subsurface processes.

Although these developments have increased the physical basis of land modeling, greater complexity has at times contributed to more differences among climate models (Gates et al. 1999). However, the advent of systematic programs comparing land models, such as the Project for Intercomparison of Land Surface Parameterization Schemes (PILPS, Henderson-Sellers et al. 1995; Henderson-Sellers 2006) has led gradually to more agreement with observations and among land models (Overgaard, Rosbjerg, and Butts 2006), in part because additional observations have been used to constrain their behavior. However, choices for adding processes and increasing realism have varied among land-surface models (e.g., Randall et al. 2007).

Figure 2.2 shows schematically the types of physical processes included in typical land models. Note that the schematic in the figure describes a land model used for both weather forecasting and climate simulation, an indication of the increasing sophistication demanded by both. The figure also hints at important biophysical and biogeochemical processes that gradually have been added and continue to be added to land models used for climate simulation, such as biophysical controls on transpiration and carbon uptake.

Figure 2.2. Schematic of Physical Processes in a Contemporary Land Model.

[Adapted from Fig. 6 in F. Chen and J. Dudhia 2001: Coupling an advanced land surface–hydrology model with the Penn State–NCAR MM5 modeling system. Part I: Model implementation and sensitivity, *Monthly Weather Review*, **129**, 569–585. Reproduced by permission of the American Meteorological Society.]

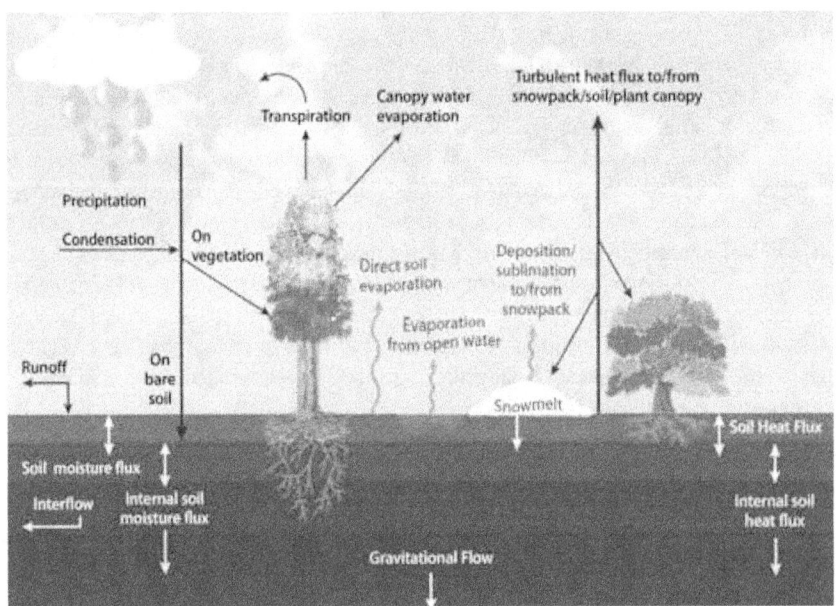

Some of the most extensive increases in complexity and sophistication have occurred with vegetation modeling in land models. An early generation of land models (Wilson et al. 1987; Sellers et al. 1986) introduced biophysical controls on plant transpiration by adding a vegetation canopy over the surface, thereby implementing vegetative control on the terrestrial water cycle. These models included exchanges of energy and moisture among the surface, canopy, and atmosphere, along with momentum loss to the surface. Further developments included improved plant physiology that allowed simulation of carbon dioxide fluxes (e.g., Bonan 1995; Sellers et al. 1996). This method lets the model treat the flow of water and carbon dioxide as an optimization problem, balancing carbon uptake for photosynthesis against water loss through transpiration. Improvements also included implementation of model parameters that could be calibrated with satellite observation (Sellers et al. 1996), thereby allowing global-scale calibration.

Continued development has included more realistic parameterization of roots (Arora and Boer 2003; Kleidon 2004) and the addition of multiple canopy layers (e.g., Gu et al. 1999; Baldocchi and Harley 1995; Wilson et al. 2003). The latter method, however, has not been used in climate models because the added complexity of multicanopy models renders unambiguous calibration very difficult. An important ongoing advance is the incorporation of biological processes that produce carbon sources and sinks through vegetation growth and decay and the cycling of carbon in the soil (e.g., Li et al. 2006), although considerable work is needed to determine observed magnitudes of carbon uptake and depletion.

Most land models assume soil with properties that correspond to inorganic soils, generally consistent with mixtures of loam, sand, and clay. High-latitude regions, however, may have extensive zones of organic soils (peat bogs), and some models have included organic soils topped by mosses, which has led to decreased soil heat flux and increased surface-sensible and latent-heat fluxes (Beringer et al. 2001).

Climate models initially treated snow as a single layer that could grow through snowfall or de-

plete though melt (e.g., Dickinson, Henderson-Sellers, and Kennedy 1993). Some recent land models for climate simulation include subgrid distributions of snow depth (Liston 2004) and blowing (Essery and Pomeroy 2004). Snow models now may use multiple layers to represent fluxes through the snow (Oleson et al. 2004). Effort also has gone into including and improving effects of soil freezing and thawing (Koren et al. 1999; Boone et al. 2000; Warrach, Mengelkamp, and Raschke 2001; Li and Koike 2003; Boisserie et al. 2006), although permafrost modeling is more limited (Malevsky-Malevich et al. 1999; Yamaguchi, Noda, and Kitoh 2005).

Vegetation interacts with snow by covering it, thereby masking snow's higher albedo (Betts and Ball 1997) and retarding spring snowmelt (Sturm et al. 2005). The net effect is to maintain warmer temperatures than would occur without vegetation masking (Bonan, Pollard, and Thompson 1992). Vegetation also traps drifting snow (Sturm et al. 2001), insulating the soil from subfreezing winter air temperatures and potentially increasing nutrient release and enhancing vegetation growth (Sturm et al. 2001). Albedo masking is included in some land-surface models, but it requires accurate simulations of snow depth to produce accurate simulation of surface-atmosphere energy exchanges (Strack, Pielke, and Adegoke 2003).

Time-evolving ice sheets and mountain glaciers are not included in most climate models. Ice sheets once were thought to be too sluggish to respond to climate change in less than a century. However, observations via satellite altimetry, synthetic aperture radar interferometry, and gravimetry all suggest rapid dynamic variability of ice sheets, possibly in response to climatic warming (Rignot and Kanagaratnam 2006; Velicogna and Wahr 2006). Most global climate models to date have been run with prescribed, immovable ice sheets. Several modeling groups are now experimenting with the incorporation of dynamic ice sheet models. Substantial physical, numerical, and computational improvements, however, are needed to provide reliable projections of 21st Century ice sheet changes. Among major challenges are incorporation of a unified treatment of stresses within ice sheets, improved methods of downscaling atmospheric

fields to the finer ice sheet grid, realistic parameterizations of surface and subglacial hydrology (fast dynamic processes controlled largely by water pressure and extent at the base of the ice sheet), and models of ice shelf interactions with ocean circulation. Ocean models, which usually assume fixed topography, may need to be modified to include flow beneath advancing and retreating ice. Meeting these challenges will require increased interaction between the glaciological and climate modeling communities, which until recently have been largely isolated from each another.

The initial focus of land models was vertical coupling of the surface with the overlying atmosphere. However, horizontal water flow through river routing has been available in some models for some time (e.g., Sausen, Schubert, and Dümenil et al. 1994; Hagemann and Dümenil 1998), with spatial resolution of routing in climate models increasing in more recent versions (Ducharne et al. 2003). Freezing soil poses additional challenges for modeling runoff (Pitman et al. 1999), with more recent work showing some skill in representing its effects (Luo et al. 2003; Rawlins et al. 2003; Niu and Yang 2006).

Work also is under way to couple groundwater models into land models (e.g., Gutowski et al. 2002; York et al. 2002; Liang, Xie, and Huang 2003; Maxwell and Miller 2005; Yeh and Eltahir 2005). Groundwater potentially introduces longer time scales of interaction in the climate system in places where it has contact with vegetation roots or emerges through the surface.

Land models encompass spatial scales ranging from model grid-box size down to biophysical and turbulence processes operating on scales the size of leaves. Explicit representation of all these scales in a climate model is beyond the scope of current computing systems and the observing systems that would be needed to provide adequate model calibration for global and regional climate. Model fluxes do not represent a single point but rather the behavior in a grid box that may be many tens or hundreds of kilometers across. Initially, these grid boxes were treated as homogeneous units but, starting with the pioneering work of Avissar and Pielke (1989), many land models have tiled a grid box

with patches of different land-use and vegetation types. Although these patches may not interact directly with their neighbors, they are linked by their coupling to the grid box's atmospheric column. This coupling does not allow for possible small-scale circulations that might occur because of differences in surface-atmosphere energy exchanges among patches (Segal and Arritt 1992; Segal et al. 1997). Under most conditions, however, the imprint of such spatial heterogeneity on the overlying atmospheric column appears to be limited to a few meters above the surface (e.g., Gutowski, Ötles, and Chen 1998).

Vertical fluxes linking the surface, canopy, and near-surface atmosphere generally assume some form of down-gradient diffusion, although counter-gradient fluxes can exist in this region much as in the overlying atmospheric boundary layer. Some attempts have been made to replace diffusion with more advanced Lagrangian random-walk approaches (Gu et al. 1999; Baldocchi and Harley 1995; Wilson et al. 2003).

Topographic variation within a grid box usually is ignored in land modeling. Nevertheless, implementing detailed river-routing schemes requires accurate digital elevation models (e.g., Hirano, Welch, and Lang 2003; Saraf et al. 2005). In addition, some soil water schemes include effects of land slope on water distribution (Choi, Kumar, and Liang 2007) and surface radiative fluxes (Zhang et al. 2006).

Validation of land models, especially globally, remains a problem due to lack of measurements for relevant quantities such as soil moisture and energy, momentum, moisture flux, and carbon flux. The PILPS project (Henderson-Sellers et al. 1995) has allowed detailed comparisons of multiple models with observations at points around the world having different climates, thus providing some constraint on the behavior of land models. Global participation in PILPS has led to more understanding of differences among schemes and improvements. Compared to previous generations, the latest land surface models exhibit relatively smaller differences from current observation-based estimates of the global distribution of surface fluxes, but the reliability of such estimates remains elusive (Henderson-Sellers et al. 2003). River routing can provide a diagnosis vs observations of a land model's spatially distributed behavior (Kattsov et al. 2000). Remote sensing has been useful for calibrating models developed to exploit it but generally has not been used for model validation. Regional observing networks that aspire to give Earth system observations, such as some mesonets in the United States, offer promise of data from spatially distributed observations of important fields for land models.

Land modeling has developed in other disciplines roughly concurrently with advances in climate models. Applications are wide ranging and include detailed models used for planning water resources (Andersson et al. 2006), managing ecosystems (e.g., Tenhunen et al. 1999), estimating crop yields (e.g., Jones and Kiniry 1986; Hoogenboom, Jones, and Boote 1992), simulating ice-sheet behavior (Peltier 2004), and projecting land use such as transportation planning (e.g., Schweitzer 2006). As suggested by this list, widely disparate applications have developed from differing scales of interest and focus. Development in some other applications has informed advances in land models for climate simulation, as in representation of vegetation and hydrologic processes. Because land models do not include all climate system features, they can be expected in future to engage other disciplines and encompass a wider range of processes, especially as resolution increases.

2.4 SEA-ICE MODELS

Most climate models include sea-ice components that have both dynamic and thermodynamic elements. That is, models include the physics governing ice movement as well as that related to heat and salt transfer within the ice. While sea ice in the real world appears as ice floes on the scale of meters, in climate models sea ice is treated as a continuum with an effective large-scale rheology describing the relationship between stress and flow.

Rheologies commonly in use are the standard Hibler viscous-plastic (VP) rheology (Hibler 1979; Zhang and Rothrock 2000) and the more-complex elastic-viscous-plastic (EVP) rheology of Hunke and Dukowicz (1997), designed primarily to improve the computational efficiency of ice models. The EVP method explicitly solves for the ice-stress tensor, while the VP solution uses an implicit iterative approach. As examples, the GFDL models (Delworth et al. 2006) and Community Climate System Model, Version 3 (CCSM3) (Collins et al. 2006a) use the EVP rheology, while the GISS models use the VP implementation.

The thermodynamic portions of sea ice models also vary. Earlier generations of climate models generally used the sea ice thermodynamics of Semtner (1976), which includes one snow layer and two ice layers with constant heat conductivities together with a simple parameterization of brine (salt) content. The GFDL climate models continue to use this but also include the interactions between brine content and heat capacity (Winton 2000). The CCSM3 and GISS models use variations (Bitz and Lipscomb 1999, Briegleb et al. 2002) incorporating additional physical processes within the ice, such as the melting of internal brine regions. Different models define snow and ice layers and ice categories differently, but all include an open water category. Typically, ice models share the grid structure of the underlying ocean model.

The albedo (proportion of incident sunlight reflected from a surface) of snow and ice plays a significant role in the climate system. Sea-ice models parameterize the albedo using expressions based on a mix of radiative transfer theory and empiricism. Figure 2.3 from Curry, Schramm, and Ebert (1995) illustrates sea-ice system interrelations and how the albedo is a function of snow or ice thickness, ice extent, open water, and surface temperature, and other factors. Models treat these factors in similar ways but vary on details. For example, the CCSM3 sea-ice component does not include dependence on solar elevation angle (Briegleb et al. 2002), but the GISS model does (Schmidt et al. 2006). Both models include the contribution of melt ponds (Ebert and Curry 1993; Schramm et al. 1997). The GFDL model follows Briegleb et al. (2002) but accounts for different effects of the different wavelengths comprising sunlight (Delworth et al. 2006).

Figure 2.3. Schematic Diagram of Sea Ice– Albedo Feedback Mechanism.

Arrow direction indicates the interaction direction. The "+" signs indicate positive interaction (i.e., increase in the first quantity leads to increase in the second quantity), and the "–" signs indicate negative interaction (i.e., increase in the first quantity leads to decrease in the second quantity). The "+/–" signs indicate either that the interaction sign is uncertain or that the sign changes over the annual cycle. [From Fig. 6 in J.A. Curry, J. Schramm, and E.E. Ebert 1995: On the sea ice albedo climate feedback mechanism, *J. Climate*, **8**, 240–247. Reproduced by permission of the American Meteorological Society.]

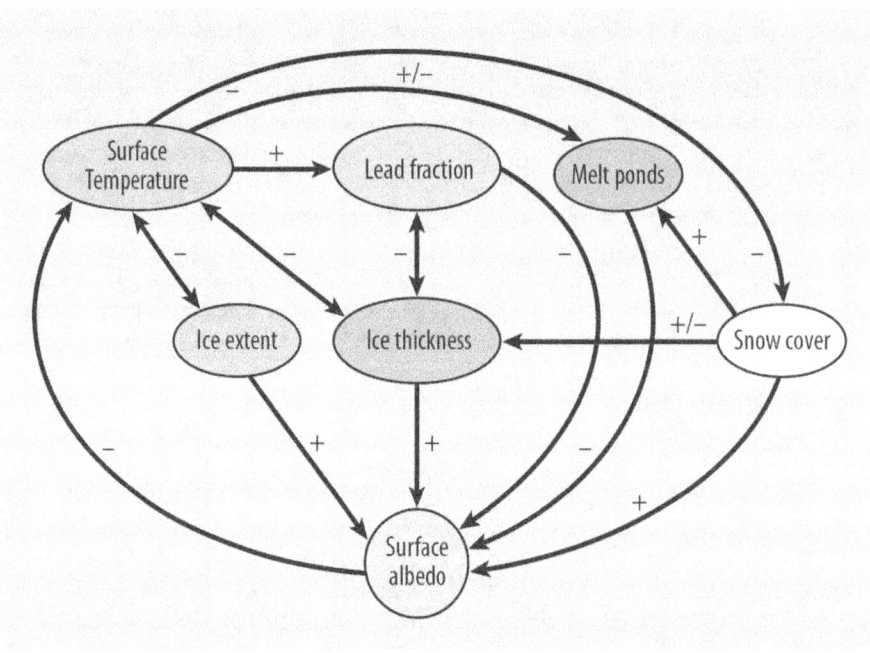

2.5 COMPONENT COUPLING AND COUPLED MODEL EVALUATION

The climate system's complexity and our inability to resolve all relevant processes in models result in a host of choices for development teams. Differing expertise, experience, and interests result in distinct pathways for each climate model. While we eventually expect to see model convergence forced by increasing insights into the climate system's workings, we are still far from that limit today in several important areas. Given this level of uncertainty, multiple modeling approaches clearly are needed. Models vary in details primarily because development teams have different ideas concerning underlying physical mechanisms relevant to the system's less-understood features. In the following, we describe some key aspects of model development by the three U.S. groups that contributed models to the IPCC Fourth Assessment (IPCC 2007). Particular focus is on points most relevant for simulating the 20th Century global mean temperature and on the model's climate sensitivity.

2.5.1 NOAA GFDL Model-Development Path

NOAA's GFDL conducted a thorough restructuring of its atmospheric and climate models for more than 5 years prior to its delivery of models to the CMIP3 database in 2004. This was done partly in response to the need for modernizing software engineering and partly in response to new ideas in modeling the atmosphere, ocean, and sea ice. Differences between the resulting models and the previous generation of climate models at GFDL are varied and substantial. Mapping out exactly why climate sensitivity and other considerations of climate simulations differ between these two generations of models would be very difficult and has not been attempted. Unlike the earlier generation, however, the new models do not use flux adjustments; some other improvements are discussed below.

The new atmospheric models developed at GFDL for global warming studies are referred to as AM2.0 and AM2.1 (GFDL Atmospheric Model Development Team 2004). Key points of departure from previous GFDL models are the adoption of a new numerical core for solving fluid dynamical equations for the atmosphere, the inclusion of liquid and ice concentrations as prognostic variables, and new parameterizations for moist convection and cloud formation. Much atmospheric development was based on running the model over observed sea-surface temperature and sea-ice boundary conditions from 1980 to 2000, with a focus on both the mean climate and the atmospheric response to El Niño–Southern Oscillation (ENSO) variability in the tropical Pacific. Given the basic model configuration, several subgrid closures were varied to optimize climate features. Modest improvements in the midlatitude wind field were obtained by adjusting the "orographic gravity wave drag," which accounts for the effects of force exerted on the atmosphere by unresolved topographic features. Substantial improvements in simulating tropical rainfall and its response to ENSO were the result of parameter optimization as well, especially the treatment of vertical transport of horizontal momentum by moist convection.

The ocean model chosen for this development is the latest version of the modular ocean model (MOM) developed over several decades at GFDL. Notable new features in this version are a grid structure better suited to simulating the Arctic Ocean and a framework for subgrid-scale mixing that avoids unphysical mixing among oceanic layers of differing densities (Gent and McWilliams 1990; Griffies 1998). A new sea-ice model includes an EVP large-scale effective rheology that has proven itself in the past decade in several models and multiple ice thicknesses in each grid box. The land model chosen is relatively simple, with vertically resolved soil temperature but retaining the "bucket hydrology" from the earlier generation of models.

The resulting climate model was studied, restructured, and tuned for an extended period, with particular interest in optimizing the structure and frequency of the model's spontaneously generated El Niño events, minimizing surface temperature biases, and maintaining an Atlantic overturning circulation of sufficient strength. During this development phase, climate sensitivity was monitored by integrating the model to equilibrium with doubled CO_2 when coupled to a "flux-adjusted" slab ocean model. A single model modification reduced the model's sensi-

tivity range from 4.0 to 4.5 K to between 2.5 and 3.0 K, as discussed further in Chapter 4. The change responsible for this reduction was inclusion of a new model of mixing in the planetary boundary near the Earth's surface. GFDL included the mixing model because it generated more-realistic boundary-layer depths and near-surface relative humidities. Sensitivity reduction resulted from modifications to the low-level cloud field; the size of this reduction was not anticipated.

Aerosol distributions used by the model were computed offline from the MOZART II model as described in Horowitz et al. (2003). No attempt was made to simulate indirect aerosol effects (interactions between clouds and aerosols), as confidence in the schemes tested was deemed insufficient. In 20th Century simulations, solar variations followed the prescription of Lean, Beer, and Bradley (1995), while volcanic forcing was based on Sato et al. (1993). Stratospheric ozone was prescribed, with the Southern Hemisphere ozone hole prescribed in particular, in 20th Century simulations. A new detailed land-use history provided a time history of vegetation types.

Final tuning of the model's global energy balance, using two parameters in the cloud prediction scheme, was conducted by examining control simulations of the fully coupled model using fixed 1860 and 1990 forcings (see box, Tuning the Global Mean Energy Balance). The resulting model is described in Delworth et al. (2006) and Gnanadesikan et al. (2006). IPCC-relevant runs of this model (CM2.0) were provided to the CMIP3–IPCC archive. Simulations of the 20th Century with time-varying forcings provided to the database and described in Knutson et al. (2006) were the first of this kind generated with this model. The model was not retuned, and no iteration of the aerosol or any other time-varying forcings followed these initial simulations.

Model development proceeded in the interim, and a new version emerged rather quickly in which the atmospheric model's numerical core was replaced by a "finite-volume" code (Lin and Rood 1996). Treatment of wind fields near the surface improved substantially, which in turn resulted in enhanced extratropical ocean

Tuning the Global Mean Energy Balance

A procedure common to all comprehensive climate models is tuning the global mean energy balance. A climate model must be in balance at top of atmosphere (TOA) and globally averaged to within a few tenths of a W/m2 in its control (pre-1860) climate if it is to avoid temperature drifts in 20th and 21st century simulations that would obscure response to imposed changes in greenhouse, aerosol, volcanic, and solar forcings. Especially because of difficulty in modeling clouds but also even in clear sky, untuned models do not currently possess this level of accuracy in their radiative fluxes. Untuned imbalances more typically range up to 5 W/m^2. Parameters in the cloud scheme are altered to create a balanced state, often taking care that individual components of this balance—the absorbed solar flux and emitted infrared flux—are individually in agreement with observations, since these help ensure the correct distribution of heating between atmosphere and ocean. This occasionally is referred to as "final tuning" the model to distinguish it from various choices made for other reasons while the model is being configured.

The need for final tuning does not preclude the use of these models for global warming simulations in which radiative forcing itself is on the order of several W/m^2. Consider, for example, the Ramaswamy et al. (2001) study on the effects of modifying the "water vapor continuum" treatment in a climate model. This is an aspect of the radiative transfer algorithm in which there is significant uncertainty. While modifying continuum treatment can change the TOA balance by more than 1 W/m^2, the effect on climate sensitivity is found to be insignificant. The change in radiative transfer in this instance alters the outgoing infrared flux by roughly 1%, and it affects the sensitivity (by changing the flux derivative with respect to temperature) by roughly the same percentage. A sensitivity change of this magnitude, say from 3 K to 3.03 K, is of little consequence given uncertainties in cloud feedbacks. The strength of temperature-dependent feedbacks, not errors in mean fluxes per se, is of particular concern in estimating climatic responses.

circulation and temperatures. ENSO variability increased in this model to unrealistically large values; however, the ocean code's efficiency also improved substantially. With retuning of the clouds for global energy balance, the new model CM2.1 was deemed to be an improved model over CM2.0 in several respects, warranting the generation of a new set of database runs. CM2.1, when run with a slab-ocean model, was found to have somewhat increased sensitivity. However, transient climate sensitivity—the global mean warming at the time of CO_2 doubling in a fully coupled model with 1% a year increase in CO_2—actually is slightly smaller than in CM2.0. Solar, aerosol, volcanic, and greenhouse gas forcings are identical in the two models.

2.5.2 Community Climate System Model-Development Path

CCSM3 was released to the climate community in June 2004. CCSM3 is a coupled climate model with components representing the atmosphere, ocean, sea ice, and land surface connected by a flux coupler. CCSM3 is designed to produce realistic simulations over a wide range of spatial resolutions, enabling inexpensive simulations lasting several millennia or detailed studies of continental-scale dynamics, variability, and climate change. Twenty-six papers documenting all aspects of CCSM3 and runs performed with it were published in a special issue of the Journal of Climate 19(11) (June 2006). The atmospheric component of CCSM3 is a spectral model. Three different resolutions of CCSM3 are supported. The highest resolution is the configuration used for climate-change simulations, with a T85 grid for atmosphere and land and a grid with around 1° resolution for ocean and sea ice but finer meridional resolution near the equator. The second resolution is a T42 grid for atmosphere and land with 1° ocean and sea-ice resolution. A lower-resolution version, designed for paleoclimate studies, has T31 resolution for atmosphere and land and a 3° version of ocean and sea ice.

The new CCSM3 version incorporates several significant improvements in physical parameterizations. Enhancements in model physics are designed to reduce several systematic biases in mean climate produced by previous CCSM versions. These enhancements include new treatments of cloud processes, aerosol radiative forcing, land-atmosphere fluxes, ocean mixed-layer processes, and sea-ice dynamics. Significant improvements are shown in sea-ice thickness, polar radiation budgets, tropical sea-surface temperatures, and cloud radiative effects. CCSM3 produces stable climate simulations of millennial duration without ad hoc adjustments to fluxes exchanged among component models. Nonetheless, there are still systematic biases in ocean-atmosphere fluxes in coastal regions west of continents, the spectrum of ENSO variability, spatial distribution of precipitation in tropical oceans, and continental precipitation and surface air temperatures. Work is under way to produce the next version of CCSM, which will reduce these biases further, and to extend CCSM to a more accurate and comprehensive model of the complete Earth climate system.

CCSM3's climate sensitivity is weakly dependent on the resolution used. Equilibrium temperature increase due to doubling carbon dioxide, using a slab-ocean model, is 2.71°C, 2.47°C, and 2.32°C, respectively, for the T85, T42, and T31 atmosphere resolutions. The transient climate temperature response to doubling carbon dioxide in fully coupled integrations is much less dependent on resolution, being 1.50°C, 1.48°C, and 1.43°C, respectively, for the T85, T42, and T31 atmosphere resolutions (Kiehl et al. 2006).

The following CCSM3 runs were submitted for evaluation for the IPCC Fourth Assessment Report and to the Program for Climate Model Diagnosis and Intercomparison (called PCMDI) for dissemination to the climate scientific community: long, present day, and 1870 control runs; an ensemble of eight 20th Century runs; and smaller ensembles of future scenario runs for the A2, A1B, and B1 scenarios and for the 20th Century commitment run where carbon dioxide levels were kept at their 2000 values. The control and 20th Century runs have been documented and analyzed in several papers in the Journal of Climate special issue, and future climate change projections using CCSM3 have been documented by Meehl et al. (2006).

2.5.3 GISS Development Path

The most recent version of the GISS atmospheric GCM, ModelE, resulted from a substantial reworking of the previous version, Model II′. Although model physics has become more complex, execution by the user is simplified as a result of modern software engineering and improved model documentation embedded within the code and accompanying web pages. The model, which can be downloaded from the GISS website by outside users, is designed to run on myriad platforms ranging from laptops to a variety of multiprocessor computers, partly because of NASA's rapidly shifting computing environment. The most recent (post-AR4) version can be run on an arbitrarily large number of processors.

Historically, GISS has eschewed flux adjustment. Nonetheless, the net energy flux at the top of atmosphere (TOA) and surface has been reduced to near zero by adjusting threshold relative humidity for water and ice cloud formation, two parameters that otherwise are weakly constrained by observations. Near-zero fluxes at these levels are necessary to minimize drift of either the ocean or the coupled climate.

To assess climate-response sensitivity to treatment of the ocean, ModelE has been coupled to a slab-ocean model with prescribed horizontal heat transport, along with two ocean GCMs. One GCM, the Russell ocean (Russell, Miller, and Rind 1995), has 13 vertical layers and horizontal resolution of 4° latitude by 5° longitude and is mass conserving (rather than volume conserving like the GFDL MOM). Alternatively, ModelE is coupled to the Hybrid Coordinate Ocean Model (HYCOM), an isopycnal model developed originally at the University of Miami (Bleck et al. 1992). HYCOM has 2° latitude by 2° longitude resolution at the equator, with latitudinal spacing decreasing poleward with the cosine of latitude. A separate rectilinear grid is used in the Arctic to avoid polar singularity and joins the spherical grid around 60°N.

Climate sensitivity to CO_2 doubling depends upon the ocean model due to differences in sea ice. Climate sensitivity is 2.7°C for the slab-ocean model and 2.9°C for the Russell ocean

GCM (Hansen et al. 2005). As at GFDL and CCSM, no effort is made to match a particular sensitivity, nor is the sensitivity or forcing adjusted to match 20th Century climate trends (Hansen et al. 2007). Aerosol forcing is calculated from prescribed concentration, computed offline by a physical model of the aerosol life cycle. In contrast to GFDL and NCAR models, ModelE includes a representation of the aerosol indirect effect. Cloud droplet formation is related empirically to the availability of cloud condensation nuclei, which depends upon the prescribed aerosol concentration (Hansen et al. 2005; Menon and Del Genio 2007).

Flexibility is emphasized in model development (Schmidt et al. 2006). ModelE is designed for a variety of applications ranging from simulation of stratospheric dynamics and middle-atmosphere response to solar forcing to projection of 21st Century trends in surface climate. Horizontal resolution typically is 4° latitude by 5° longitude, although twice that resolution is used more often for studies of cloud processes. The model top has been raised from 10 mb (as in the previous Model II ') to 0.1 mb, so the top has less influence on stratospheric circulation. Coding emphasizes "plug-and-play" structure, so the model can be adapted easily for future needs such as fully interactive carbon and nitrogen cycles.

Model development is devoted to improving the realism of individual model parameterizations, such as the planetary boundary layer or sea-ice dynamics. Because of the variety of applications, relatively little emphasis is placed on optimizing the simulation of specific phenomena such as El Niño or the Atlantic thermohaline circulation; as noted above, successful reproduction of one phenomenon usually results in a suboptimal simulation of another. Nonetheless, some effort was made to reduce biases in previous model versions that emerged from the interaction of various model features such as subtropical low clouds, tropical rainfall, and variability of stratospheric winds. Some model adjustments were structural, as opposed to the adjustment of a particular parameter—for example, introduction of a new planetary boundary layer parameterization that reduced unrealistic cloud formation in the lowest model level (Schmidt et al. 2006).

Because of their uniform horizontal coverage, satellite retrievals are emphasized for model evaluation like Earth Radiation Budget Experiment fluxes at TOA, Microwave Sounding Unit channels 2 (troposphere) and 4 (stratosphere) temperatures, and International Satellite Cloud Climatology Project (ISCCP) diagnostics. Comparison to ISCCP is through a special algorithm that samples GCM output to mimic data collection by an orbiting satellite. For example, high clouds may include contributions from lower levels in both the model and the downward-looking satellite instrument. This satellite perspective within the model allows a rigorous comparison to observations. In addition to satellite retrievals, some GCM fields like zonal wind are compared to in situ observations adjusted by European Center for Medium Range Weather Forecasts' 40-year reanalysis data (Uppala et al. 2005). Surface air temperature is taken from the Climate Research Unit gridded global surface temperature dataset (Jones et al. 1999).

2.5.4 Common Problems

The CCSM and GFDL development teams met several times to compare experiences and discuss common problems in the two models. A subject of considerable discussion and concern was the tendency for an overly strong "cold tongue" to develop in the eastern equatorial Pacific Ocean and for associated errors to appear in the pattern of precipitation in the Inter-Tropical Convergence Zone (often referred to as the "double-ITCZ problem"). Meeting attendees noted that the equilibrium climate sensitivities of the two models to doubled atmospheric carbon dioxide (see Chapter 4) had converged from earlier generations in which the NCAR model was on the low end of the canonical sensitivity range of 1.5 to 4.5 K, while the GFDL model was near the high end. This convergence in global mean sensitivity was considered coincidental because no specific actions were taken to engineer convergence. It was not accompanied by any noticeable convergence in cloud-feedback specifics or in the regional temperature changes that make up global mean values.

2.6 REDUCTIVE VS HOLISTIC EVALUATION OF MODELS

To evaluate models, appreciation of their structure is necessary. For example, discussion of climatic response to increasing greenhouse gases is intimately related to the question of how infrared radiation escaping to space is controlled. When summarizing results from climate models, modelers often speak and think in terms of a simple energy balance model in which the global mean infrared energy escaping to space has a simple dependence on global mean surface temperature. Water vapor or cloud feedbacks often are incorporated into such global mean energy balance models with simple relationships that can be tailored easily to generate a desired result. In contrast, Fig. 2.4 shows a snapshot at an instant when infrared radiation is escaping to space in the kind of AGCM discussed in this report. Detailed distributions of clouds and water vapor simulated by the model and transported by the model's evolving wind fields create complex patterns in space and time that, if the simulation is sufficiently realistic, resemble images seen from satellites viewing Earth at infrared wavelengths.

As described above, AGCMs evolve the state of atmosphere and land system forward in time,

starting from some initial condition. They consist of rules that generate the state of a variable (e.g., temperature, wind, water vapor, clouds, rainfall rate, water storage in the land, and land-surface temperature) from its preceding state roughly a half-hour earlier. By this process a model simulates the weather over the Earth. To change the way the model's infrared radiation reacts to increasing temperatures, the rules would have to be modified.

One goal of climate modeling is to decrease empiricism and base models as much as possible on well-established physical principles. This goal is pursued primarily by decomposing the climate system into a number of relatively simple processes and interactions. Modelers focus on rules governing the evolution of these individual processes rather than working with more holistic concepts such as global mean infrared radiation escaping to space, average summertime rainfall over Africa, and average wintertime surface pressure over the Arctic. These are all outcomes of the model, determined by the set of reductive rules that govern the model's evolution.

Suppose the topic under study is how ocean temperatures affect rainfall over Africa. An empirical statistical model could be developed

Figure 2.4. A Snapshot in Time of Infrared Radiation Escaping to Space in a Version of Atmospheric Model AM2 Constructed at NOAA's Geophysical Fluid Dynamics Laboratory (GFDL 2004).

The largest amount of energy emitted is in the darkest areas, and the least is in the brightest areas. This version of the atmospheric model has higher resolution than that used for simulations in the CMIP3 archive (50 km rather than 200 km), but, other than resolution, it uses the same numerical algorithm. The resolution is typical in many current studies with atmosphere-only simulations.

using observations and standard statistical techniques in which the model is tuned to these observations. Alternatively, one can use an AGCM giving results like those pictured in Fig. 2.4. An AGCM does not deal directly with high-level climate output such as African rainfall averaged over some period. Rather, it attempts to simulate the climate system's inner workings or dynamics at a much finer level of granularity. To the extent that the simulation is successful and convincing, the model can be analyzed and manipulated to uncover the detailed physical mechanisms underlying the connection between ocean temperatures and rainfall over Africa. The AGCM-simulated connection may or may not be as good as the fit obtained with the explicitly tuned statistical model, but a reductive model ideally provides a different level of confidence in its explanatory and predictive power. See, for example, Hoerling et al. (2006) for an analysis of African rainfall and ocean temperature relationships in a set of AGCMs.

Our confidence in the explanatory and predictive power of climate models grows with their ability to simulate many climate system features simultaneously with the same set of physically based rules. When a model's ability to simulate the evolution of global mean temperature over the 20th Century is evaluated, it is important to try to make this evaluation in the context of the model's ability to spontaneously generate El Niño variability of the correct frequency and spatial structure, for example, and to capture the effects of El Niño on rainfall and clouds. Simulation quality adds confidence in the reductive rules being used to generate simultaneous simulation of all these phenomena.

A difficulty to which we will return frequently in this report is that of relating climate-simulation qualities to a level of confidence in the model's ability to predict climate change.

2.7 USE OF MODEL METRICS

Recently, objective evaluation has exploded with the wide availability of model simulation results in the CMIP3 database (Meehl et al. 2006). One important area of research is in the design of metrics to test the ability of models to simulate well-observed climate features (Reichler and Kim 2008; Gleckler, Taylor, and Dou-

triaux 2008). Aspects of observed climate that must be simulated to ensure reliable future predictions are unclear. For example, models that simulate the most realistic present-day temperatures for North America may not generate the most reliable projections of future temperature changes. Projected climate changes in North America may depend strongly on temperature changes in the tropical Pacific Ocean and the manner in which the jet stream responds to them. The quality of a model's simulation of air-sea coupling over the Pacific might be a more relevant metric. However, metrics can provide guidance about overall strengths and weaknesses of individual models, as well as the general state of modeling.

The use of metrics also can explain why the "best" climate model cannot be chosen at this time. In Fig. 2.5 below, each colored triangle represents a different metric for which each model was evaluated (e.g., "ts" represents surface temperature). The figure displays the relative error value for a variety of metrics for each model, represented by a vertical column above each tick mark on the horizontal axis. Values less than zero represent a better-than-average simulation of a particular field measured by the metric, while values greater than zero show models with errors greater than the average. The black triangles connected by the dashed line represent the normalized sum from the errors of all 23 fields. The models were ranked from left to right based on the value of this total error. As can be seen, models with the lowest total errors tend to score better than average in most individual metrics but not in all. For an individual application, the model with the lowest total errors may not be the best choice.

2.8 CLIMATE SIMULATIONS DISCUSSED IN THIS REPORT

Three types of climate simulation discussed in this report are described below. They differ according to which climate-forcing factors are used as model input.

Control runs use constant forcing. The sun's energy output and the atmospheric concentrations of carbon dioxide and other gases and aerosols do not change in control runs. As with other types of climate simulation, day-night and

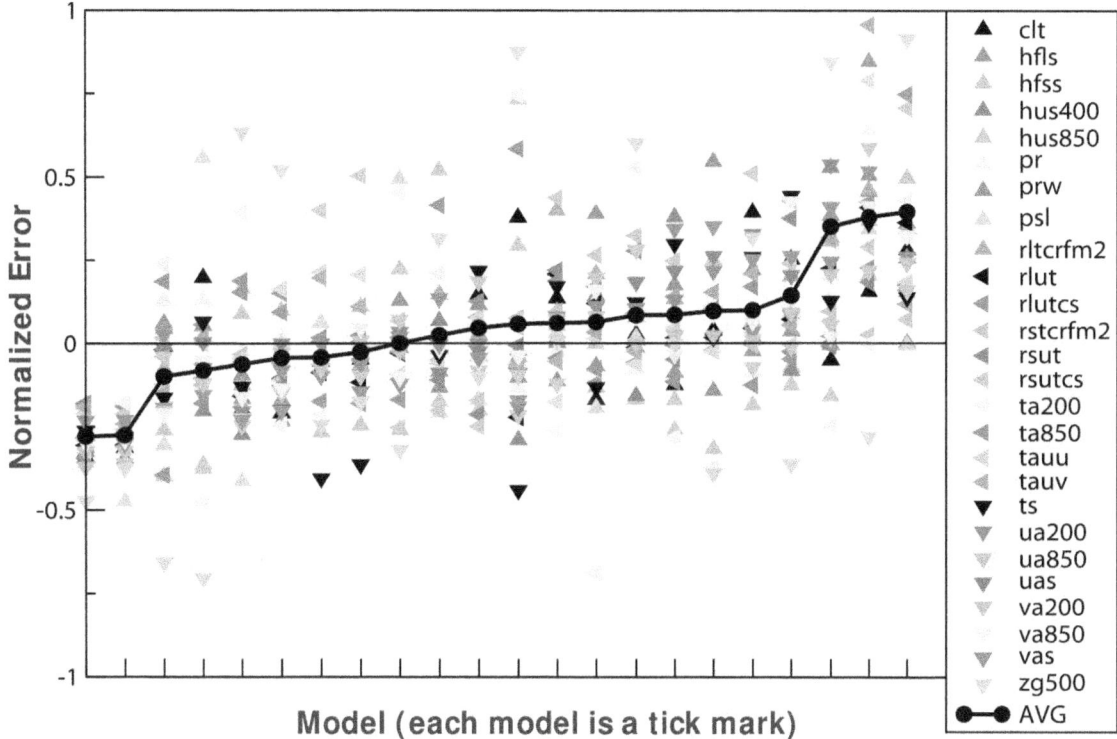

Figure 2.5. Model Metrics for 23 Different Climate Fields.

Values less than 0 indicate an error less than the average CMIP3 model, while values greater than 0 are more than the average. The black triangles connected by the black line show a total score obtained by averaging all 23 fields. Each tick mark represents a different model. [Figure adapted from P.J. Gleckler, K.E. Taylor, and C. Doutriaux 2008: Performance metrics for climate models. *J. Geophysical Research*, **113**, D06104, doi:10.1029/2007JD008972. Reproduced by permission of the American Geophysical Union (AGU).]

seasonal variations occur, along with internal "oscillations" such as ENSO. Other than these variations, the control run of a well-behaved climate model is expected eventually to reach a steady state.

Values of control-run forcing factors often are set to match present-day conditions, and model output is compared with present-day observations. Actually, today's climate is affected not only by current forcing but also by the history of forcing over time—in particular, past emissions of greenhouse gases. Nevertheless, present-day control-run output and present-day observations are expected to agree fairly closely if models are reasonably accurate. We compare model control runs with observations in Chapter 5.

Idealized climate simulations are aimed at understanding important processes in models and in the real world. They include experiments in which the amount of atmospheric carbon dioxide increases at precisely 1% per year (about twice the current rate) or doubles instantaneously. Carbon dioxide doubling experiments typically are run until the simulated climate reaches a steady state of equilibrium with the enhanced greenhouse effect. Until the mid-1990s, idealized simulations often were employed to assess possible future climate changes including human-induced global warming. Recently, however, more realistic time-evolving simulations (defined immediately below) have been used for making climate predictions. We discuss idealized simulations and their implications for climate sensitivity in Chapter 4.

Time-dependent climate-forcing simulations are the most realistic, especially for eras in which climate forcing is changing rapidly, such as the 20th and 21st centuries. Input for 20th Century simulations includes observed time-varying values of solar energy output, atmospheric carbon dioxide, and other climate-relevant gases and aerosols, including those produced in volcanic eruptions. Each modeling group uses its own best estimate of these factors. Significant uncertainties occur in many of them, especially atmospheric aerosols, so different models use different input for their 20th Century simulations. We discuss uncertainties in climate-forcing factors in Chapter 4 and 20th Century simulations in Chapter 5 after comparing control runs with observations.

Time-evolving climate forcing also is used as input for modeling future climate change. This subject is discussed in CCSP Synthesis and Assessment Product 3.2. Finally, we mention for the record simulations of the distant past (various time periods ranging from early Earth up to the 19th Century). These simulations are not discussed in this report, but some of them have been used to loosely "paleocalibrate" simulations of the more recent past and the future (Hoffert and Covey 1992; Hansen et al. 2006; Hegerl et al. 2006).

Added Value of
Regional Climate
Model Simulations

3.1 TYPES OF DOWNSCALING SIMULATIONS

This section focuses on downscaling using three-dimensional models based on fundamental conservation laws [i.e., numerical models with foundations similar to general circulation models (GCMs)]. A later section of the chapter discusses an alternative method, statistical downscaling.

There are three primary approaches to numerical downscaling:

- Limited-area models (Giorgi and Mearns 1991, 1999; McGregor 1997; Wang et al. 2004).

- Stretched-grid models (e.g., Déqué and Piedelievre 1995; Fox-Rabinovitz et al. 2001, 2006).

- Uniformly high resolution atmospheric GCMs (AGCMs) (e.g., Brankovic and Gregory 2001; May and Roeckner 2001; Duffy et al. 2003; Coppola and Giorgi 2005).

Limited-area models, also known as regional climate models (RCMs), have the most widespread use. The third method sometimes is called "time-slice" climate simulation because the AGCM simulates a portion of the period represented by the coarser-resolution parent GCM that supplies the model's boundary conditions. All three methods use interactive land models, but sea-surface temperatures and sea ice generally are specified from observations or an atmosphere-ocean GCM (AOGCM). All three also are used for purposes beyond downscaling global simulations, most especially for studying climatic processes and interactions on scales too fine for typical GCM resolutions.

As limited-area models, RCMs cover only a portion of the planet, typically a continental domain or smaller. They require lateral boundary conditions (LBCs), obtained from observations such as atmospheric analyses (e.g., Kanamitsu et al. 2002; Uppala et al. 2005) or a global simulation. There has been limited two-way coupling wherein an RCM supplies part of its output back to the parent GCM (Lorenz and Jacob 2005). Simulations with observation-based boundary conditions are used not only to study fine-scale climatic behavior but also to help segregate GCM errors from those intrinsic to the RCM when performing climate change simulations (Pan et al. 2001). RCMs also may use grids nested inside a coarser RCM simulation to achieve higher resolution in subregions (e.g., Liang, Kunkel, and Samel 2001; Hay et al. 2006).

Stretched-grid models, like high-resolution AGCMs, are global simulations but with spatial resolution varying horizontally. The highest resolution may focus on one (e.g., Déqué and Piedelievre 1995; Hope, Nicholls, and McGregor 2004) or a few regions (e.g., Fox-Rabinovitz, Takacs, and Govindaraju 2002). In some sense, the uniformly high resolution AGCMs are the upper limit of stretched-grid simulations in which the grid is uniformly high everywhere.

Highest spatial resolutions are most often several tens of kilometers, although some (e.g., Grell et al. 2000a, b; Hay et al. 2006) have simulated climate with resolutions as small as a few kilometers using multiple nested grids. Duffy et al. (2003) have performed multiple AGCM time-slice computations using the same model to simulate resolutions from 310 km down to 55 km. Higher resolution generally yields improved climate simulation, especially for fields such as precipitation that have high spatial variability. For example, some studies show that higher resolution does not have a statistically significant advantage in simulating large-scale circulation patterns but does yield better monsoon precipitation forecasts and interannual variability (Mo et al. 2005) and precipitation intensity (Roads, Chen, and Kanamitsu 2003).

Improvement in results, however, is not guaranteed: Hay et al. (2006) find deteriorating timing and intensity of simulated precipitation vs observations in their inner, high-resolution nests, even though the inner nest improves topography resolution. Extratropical storm tracks in a time-slice AGCM may shift poleward relative to the coarser parent GCM (Stratton 1999; Roeckner et al. 2006) or to lower-resolution versions of the same AGCM (Brankovic and Gregory 2001); thus these AGCMs yield an altered climate with the same sea-surface temperature distribution as the parent model.

Spatial resolution affects the length of simulation periods because higher resolution requires shorter time steps for numerical stability and accuracy. Required time steps scale with the inverse of resolution and can be much smaller than AOGCM time steps. Increases in resolution most often are applied to both horizontal directions, meaning that computational demand varies inversely with the cube of resolution. Several RCM simulations have lasted 20 to 30 years (Christensen, Carter, and Giorgi 2002;

Leung et al. 2004; Plummer et al. 2006) and even as long as 140 years (McGregor 1999) with no serious drift away from reality. Even so, the RCM, stretched-grid, and time-slice AGCM simulations typically last only months to a few years. Vertical resolution usually does not change with horizontal resolution, although Lindzen and Fox-Rabinovitz (1989) and Fox-Rabinovitz and Lindzen (1993) have expressed concerns about the adequacy of vertical resolution relative to horizontal resolution in climate models.

Higher resolution in RCMs and stretched-grid models also must satisfy numerical constraints. Stretched-grid models whose ratio of coarsest-to-finest resolution exceeds a factor of roughly 3 are likely to produce inaccurate simulation due to truncation error (Qian, Giorgi, and Fox-Rabinovitz 1999). Similarly, RCMs will suffer from incompletely simulated energy spectra and thus loss of accuracy if their resolution is about 12 times or more finer than the resolution of the LBC source, which may be coarser RCM grids (Denis et al. 2002; Denis, Laprise, and Caya 2003; Antic et al. 2004, 2006; Dimitrijevic and Laprise 2005). In addition, these same studies indicate that LBCs should be updated more frequently than twice per day.

Additional factors also govern ingestion of LBCs by RCMs. LBCs are most often ingested in RCMs by damping the model's state toward LBC fields in a buffer zone surrounding the domain of interest (Davies 1976; Davies and Turner 1977). If the buffer zone is only a few grid points wide, the interior region may suffer phase errors in simulating synoptic-scale waves (storm systems), with resulting error in the overall regional simulation (Giorgi, Marinucci, and Bates 1993). Spurious reflections also may occur in boundary regions (e.g., Miguez-Macho, Stenchikov, and Robock 2005). RCM boundaries should be where the driving data are of optimum accuracy (Liang, Kunkel, and Samel 2001), but placing the buffer zone in a region of rapidly varying topography can induce surface-pressure errors. These errors result from mismatch between the smooth topography implicit in the coarse resolution driving the data and the varying topography resolved by the model (Hong and Juang 1998). Domain size also may influence RCM results. If a domain is

too large, the model's interior flow may drift from the large-scale flow of the driving dataset (Jones, Murphy, and Noguer 1995). However, too small a domain overly constrains interior dynamics, preventing the model from generating appropriate response to interior mesoscale-circulation and surface conditions (Seth and Giorgi 1998). RCMs appear to perform well for domains roughly the size of the contiguous United States. Figure 3.1 shows that the daily, root-mean-square difference (RMSD) between simulated and observed (reanalysis) 500-hPa heights generally is within observational noise levels (about 20 m).

Because simulations from the downscaling models may be analyzed for periods as short as a month, model spinup is important (e.g., Giorgi and Bi 2000). During spinup, the model evolves to conditions representative of its own climatology, which may differ from the sources of initial conditions. The atmosphere spins up in a matter of days, so the key factor is spinup of soil moisture and temperature, which evolve more

slowly. Equally important, data for initial conditions often are lacking or have low spatial resolution, so initial conditions may be only a poor approximation of the model's climatology. Spinup is especially relevant for downscaling because these models presumably are resolving finer surface features than coarser models, with the expectation that the downscaling models are providing added value through proper representation of these surface features. Deep-soil temperature and moisture, at depths of 1 to 2 meters, may require several years of spinup. However, these deep layers generally interact weakly with the rest of the model, so shorter spinup times are used. For multiyear simulations, a period of 3 to 4 years appears to be the minimal requirement (Christensen 1999; Roads et al. 1999). This ensures that the upper meter of soil has a climatology in further simulations that is consistent with the evolving atmosphere.

Many downscaling simulations, especially with RCMs, are for periods much shorter than 2 years. Such simulations probably will not use

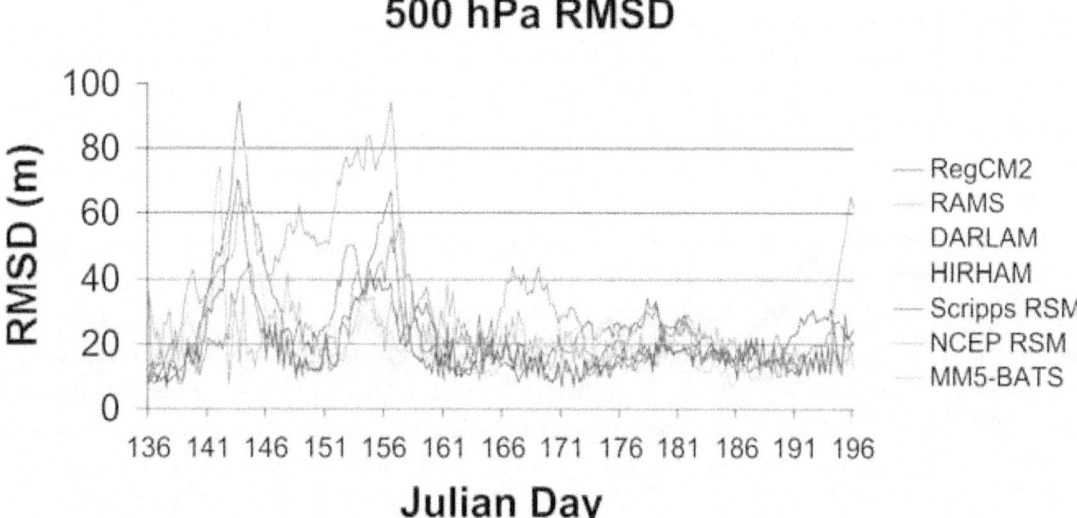

Figure 3.1. Daily Root-Mean-Square Differences (RMSD) in 500-hPa Heights Between Observations (Reanalysis) and Seven Models Participating in the PIRCS 1a Experiment, for May 15 to July 15, 1988.

RMSD values were averaged over the simulation domain inside the boundary-forcing zone. [Adapted from Fig. 4 in E.S. Takle et al. 1999: Project to Intercompare Regional Climate Simulations (PIRCS): Description and initial results. *J. Geophysical Research*, **104**, 19443–19461. Used with permission of the American Geophysical Union.]

multiyear spinup. Rather, these studies may focus on more rapidly evolving atmospheric behavior governed by LBCs, including extreme periods such as drought (Takle et al. 1999) or flood (Giorgi et al. 1996; Liang, Kunkel, and Samel 2001; Anderson, C. J., et al. 2003). Thus, they assume that interaction with the surface, while not negligible, is not strong enough to skew the atmospheric behavior studied. Alternatively, relatively short regional simulations may specify, for sensitivity study, substantial changes in surface evaporation (e.g., Paegle, Mo, and Nogués-Paegle 1996), soil moisture (e.g., Xue et al. 2001), or horizontal moisture flux at lateral boundaries (e.g., Qian, Tao, and Lau 2004).

3.1.1 Parameterization Issues

Even with higher resolution than standard GCMs, models simulating regional climate still need parameterizations for subgrid-scale processes, most notably boundary-layer dynamics, surface-atmosphere coupling, radiative transfer, and cloud microphysics. Most regional simulations also require a convection parameterization, although a few have used sufficiently fine grid spacing (a few kilometers) to allow acceptable simulation without it (e.g., Grell et al. 2000). Often, these parameterizations are the same or nearly the same as those used in GCMs. All parameterizations, however, make assumptions that they are representing the statistics of subgrid processes. Implicitly or explicitly, they require that the grid box area in the real world has sufficient samples to justify stochastic modeling. For some parameterizations such as convection, this assumption becomes doubtful when grid boxes are only a few kilometers in size (Emanuel 1994).

In addition, models simulating regional climate may include circulation characteristics, such as rapid mesoscale circulations (jets) whose interaction with subgrid processes like convection and cloud cover differs from larger-scale circulations resolved by typical GCMs. This factor is part of a larger issue, that parameterizations may have regime dependence, performing better for some conditions than for others. For example, the Grell (1993) convection scheme is responsive to large-scale tropospheric forcing,

whereas the Kain and Fritsch (1993) scheme is heavily influenced by boundary-layer forcing. As a result, the Grell scheme better simulates the propagation of precipitation over the U.S. Great Plains that is controlled by large-scale tropospheric forcing, while the Kain–Fritsch scheme better simulates late-afternoon convection peaks in the southeastern United States that are governed by boundary-layer processes (Liang et al. 2004). As a consequence, parameterizations for regional simulation may differ from their GCM counterparts, especially for convection and cloud microphysics. As noted earlier, regional simulation in some cases may have resolution of only a few kilometers, and the convection parameterization may be discarded (Grell et al. 2000). A variety of parameterizations exist for each subgrid process, with multiple choices often available in a single model (e.g., Grell, Dudhia, and Stanfler 1994; Skamarock et al. 2005).

3.1.2 Regional Simulation vs Computational Costs

The chief reason for performing regional simulation, whether by an RCM, a stretched-grid model, or a time-slice AGCM, is to resolve behavior considered important for a region's climate that a global model does not resolve. Thus, regional simulation should have clearly defined regional-scale (mesoscale) phenomena targeted for simulation. These include tropical storms (e.g., Oouchi et al. 2006), effects of mountains (e.g., Leung and Wigmosta 1999; Grell et al. 2000; Zhu and Liang 2007), jet circulations (e.g., Takle et al. 1999; Anderson et al. 2001; Anderson, C. J., et al. 2003; Byerle and Paegle 2003; Pan et al. 2004), and regional ocean-land interaction (e.g., Kim et al. 2005; Diffenbaugh, Snyder, and Sloan 2004). The most immediate value of regional simulation, then, is to explore how such phenomena operate in the climate system, an understanding of which becomes a justification for the expense of performing regional simulation. Phenomena and computational costs together influence the design of regional simulations. Simulation periods and resolution are balanced between sufficient length and number of simulations for climate statistics vs computational cost. For RCMs and stretched-grid models, the sizes of regions targeted for

high-resolution simulation are determined in part by where the phenomenon occurs.

In the context of downscaling, regional simulation offers the potential to include phenomena affecting regional climate change that are not explicitly resolved in the global simulation. When incorporating boundary conditions corresponding to future climate, regional simulation can then indicate how these phenomena contribute to climate change. Results, of course, are dependent on the quality of the boundary-condition source (Pan et al. 2001; de Elía, Laprise, and Denis 2002), although use of multiple sources of future climate may lessen this vulnerability and offer opportunity for probabilistic estimates of regional climate change (Raisanen and Palmer 2001; Giorgi and Mearns 2003; Tebaldi et al. 2005). Results also depend on the physical parameterizations used in the simulation (Yang and Arritt 2002; Vidale et al. 2003; Déqué et al. 2005; Liang et al. 2006).

Advances in computing power suggest that typical GCMs eventually will operate at resolutions of most current regional simulations (a few tens of kilometers), so understanding and modeling improvements gained for regional simulation can promote appropriate adaptation of GCMs to higher resolution. For example, interaction between mesoscale jets and convection appears to require parameterized representation of convective downdrafts and their influence on the jets (Anderson, Arritt, and Kain 2007), parameterized behavior not required for resolutions that do not resolve mesoscale circulations.

Because of the variety of numerical techniques and parameterizations employed in regional simulation, many models and versions of models exist. Generally in side-by-side comparisons (e.g., Takle et al. 1999; Anderson, C. J., et al. 2003; Fu et al. 2005; Frei et al. 2006; Rinke et al. 2006), no single model appears best vs observations, with different models showing superior performance depending on the field examined. Indeed, the best results for downscaling climate simulations and estimating climate-change uncertainty may come from assessing an ensemble of simulations (Giorgi and Bi 2000; Yang and Arritt 2002; Vidale et al. 2003; Déqué et al. 2005). Such an ensemble

may capture much of the uncertainty in climate simulation, offering an opportunity for physically based analysis of climate changes and also the uncertainty of the changes. Several regional models have performed simulations of climate change for parts of North America, but at present no regional projections have used an ensemble of regional models to simulate the same time periods with the same boundary conditions. Such systematic evaluation has occurred in Europe in the PRUDENCE (Christensen, Carter, and Giorgi 2002) and ENSEMBLES (Hewitt and Griggs 2004) projects and is starting in North America with the North American Regional Climate Change Assessment Program (NARCCAP 2007).

3.2 EMPIRICAL DOWNSCALING

Empirical or statistical downscaling is an alternative approach to obtaining regional-scale climate information (Kattenberg et al. 1996; Hewitson and Crane 1996; Giorgi et al. 2001; Wilby et al. 2004, and references therein). It uses statistical relationships to link resolved behavior in GCMs with climate in a targeted area. The targeted area's size can be as small as a single point. As long as significant statistical relationships occur, empirical downscaling can yield regional information for any desired variable such as precipitation and temperature, as well as variables not typically simulated in climate models, such as zooplankton populations (Heyen, Fock, and Greve 1998) and initiation of flowering (Maak and von Storch 1997). This approach encompasses a range of statistical techniques from simple linear regression (e.g., Wilby et al. 2000) to more-complex applications such as those based on weather generators (Wilks and Wilby 1999), canonical correlation analysis (e.g., von Storch, Zorita, and Cubasch 1993), or artificial neural networks (e.g., Crane and Hewitson 1998). Empirical downscaling can be very inexpensive compared to numerical simulation when applied to just a few locations or when simple techniques are used. Lower costs, together with flexibility in targeted variables, have led to a wide variety of applications for assessing impacts of climate change.

Some methods have been compared side by side (Wilby and Wigley 1997; Wilby et al. 1998;

Zorita and von Storch 1999; Widman, Bretherton, and Salathe 2003). These studies have tended to show fairly good performance of relatively simple vs more-complex techniques and to highlight the importance of including moisture and circulation variables when assessing climate change. Statistical downscaling and regional climate simulation also have been compared (Kidson and Thompson 1998; Mearns et al. 1999; Wilby et al. 2000; Hellstrom et al. 2001; Wood et al. 2004; Haylock et al. 2006), with no approach distinctly better or worse than any other. Statistical methods, though computationally efficient, are highly dependent on the accuracy of regional temperature, humidity, and circulation patterns produced by their parent global models. In contrast, regional climate simulation, though computationally more demanding, can improve the physical realism of simulated regional climate through higher resolution and better representation of important regional processes. The strengths and weaknesses of statistical downscaling and regional modeling thus are complementary.

3.3 STRENGTHS AND LIMITATIONS OF REGIONAL MODELS

We focus here on numerical models simulating regional climate but do not discuss empirical downscaling because the wide range of applications using the latter makes difficult a general assessment of strengths and limitations.

The higher resolution in regional-scale simulations provides quantitative value to climate simulation. With finer resolution, scientists can resolve mesoscale phenomena contributing to intense precipitation, such as stronger upward motions (Jones, Murphy, and Noguer 1995) and coupling between regional circulations and convection (e.g., Anderson, Arritt, and Kain 2007). Time-slice AGCMs show intensified storm tracks relative to their parent model (Solman, Nunez, and Rowntree 2003; Roeckner et al. 2006). Thus, although regional models may still miss the most extreme precipitation (Gutowski et al. 2003, 2007a), they can give more intense events that will be smoothed in coarser-resolution GCMs. The higher resolution also includes other types of scale-dependent variability, especially short-term variability such as extreme winds and locally extreme temperature that coarser-resolution models will smooth and thus inhibit.

Mean fields also appear to be simulated somewhat better on average than are those in coarser GCMs because spatial variations potentially are better resolved. Thus, Giorgi et al. (2001) report typical errors in RCMs of less than 2°C temperature and 50% for precipitation in regions 105 to 106 km2. Large-scale circulation fields tend to be well simulated, at least in the extratropics.

As alluded to above, regional-scale simulations also have phenomenological value, simulating processes that GCMs either cannot resolve or can resolve only poorly. These include internal circulation features such as the nocturnal jet that imports substantial moisture to the center of the United States and couples with convection (e.g., Byerle and Paegle 2003; Anderson, Arritt, and Kain 2007). These processes often have substantial diurnal variation and thus are important to proper simulation of regional diurnal cycles of energy fluxes and precipitation. Some processes require the resolution of surface features too coarse for typical GCM resolution. These include rapid topographic variation and its influence on precipitation (e.g., Leung and Wigmosta 1999; Hay et al. 2006) and the climatic influences of bodies of water such as the Gulf of California (e.g., Anderson et al. 2001) and the North American Great Lakes (Lofgren 2004) and their downstream influences. In addition, regional simulations resolve land-surface features that may be important for climate-change impact assessments such as distributions of crops and other vegetation (Mearns 2003; Mearns et al. 2003), although care is needed to obtain useful information at higher resolution (Adams, McCarl, and Mearns 2003).

An important limitation for regional simulations is that they are dependent on boundary conditions supplied from some other source. This applies to all three forms of numerical simulation (RCMs, stretched-grid models, and time-slice AGCMs), since they all typically require input of sea-surface temperature and ocean ice. Some RCM simulations have been coupled to a regional ocean-ice model, with mixed-layer ocean

(Lynch et al. 1995; Lynch, Maslanic, and Wu 2001) and a regional ocean-circulation model (Rummukainen et al. 2004), but this is not common. In addition, of course, RCMs require LBCs. Thus, regional simulations by these models are dependent on the model quality or on observations supplying boundary conditions. This is especially true for projections of future climate, suggesting value in performing an ensemble of simulations using multiple atmosphere-ocean global models to supply boundary conditions, thus including some of the uncertainty involved in constructing climate models and projecting future changes in boundary conditions.

Careful evaluation also is necessary to show differences, if any, between the regional simulation's large-scale circulation and its driving dataset. Generally, any tendency for the regional simulation to alter biases in the parent GCM's large-scale circulation should be viewed with caution (Jones, Murphy, and Noguer 1995). An RCM normally should not be expected to correct large-scale circulation problems of the parent model unless the physical basis for the improvement is clearly understood. Clear physical reasons for the correction due to higher resolution, such as better rendition of physical processes like topographic circulation (e.g., Leung and Qian 2003), surface-atmosphere interaction (Han and Roads 2004), and convection (Liang et al. 2006) must be established. Otherwise, the regional simulation may simply have errors that counteract the parent GCM's errors, thus undermining confidence in projected future climate.

RCMs also may exhibit difficulty in outflow regions of domains, especially regions with relatively strong cross-boundary flow, which may occur in extratropical domains covering a single continent or less. The difficulty appears to arise because storm systems may track across the RCM's domain at a different speed from their movement in the driving-data source, resulting in a mismatch of circulations at boundaries where storms would be moving out of the domain. Also, unresolved scales of behavior are always present, so regional simulations are still dependent on parameterization quality for the scales explicitly resolved. Finally, higher computational demand due to shorter time steps limits the length of typical simulations to 2 to 3 decades or less (e.g., Christensen, Carter, and Giorgi 2002; NARCCAP 2007), with few ensemble simulations to date.

Model Climate Sensitivity

The response of climate to a perturbation such as a change in carbon dioxide concentration, or in the flux of energy from the sun, can be divided into two factors: "radiative forcing" due to the perturbation in question and "climate sensitivity," characterizing the response of the climate per unit change in radiative forcing. Climate response is then the product of radiative forcing and climate sensitivity. This distinction is useful because of two approximations: radiative forcing often can be thought of as independent of the resulting climate response, and climate sensitivity can often be thought of as independent of the agent responsible for perturbation to the energy balance. When two or more perturbations are present simultaneously, their cumulative effect can be approximated by adding their respective radiative forcings (Hansen et al. 2006).

Climate sensitivity as traditionally defined refers to the global mean temperature, but a model's global mean temperature response is very relevant to its regional temperature responses as well. This "pattern scaling" effect is discussed at the end of this chapter.

Radiative forcing typically is calculated by changing the atmospheric composition or external forcing and computing the net trapping of heat that occurs before the climate system has had time to adjust.[1] These direct heat-trapping properties are well characterized for the most significant greenhouse gases. As a result, uncertainty in climate responses to greenhouse gases typically is dominated by uncertainties in climate sensitivity rather than in radiative forcing (Ramaswamy et al. 2001). For example, suddenly doubling the atmospheric amount of carbon dioxide would add energy to the surface and the troposphere at the rate of about 4 W/m^2 for the first few months after the doubling (Forster et al. 2007). Eventually, lower tropospheric temperatures would increase (and climate would change in other ways) in response to this forcing, Earth would radiate more energy to space, and the imbalance would diminish as the system returned to equilibrium.

4.1 CHARACTERIZING CLIMATE RESPONSE

4.1.1 Equilibrium Sensitivity and Transient Climate Response

The idea of characterizing climate response using a single number represented by climate sensitivity appeared early in the development of

---

[1] Because the stratosphere cools rapidly in response to increasing carbon dioxide and this cooling affects the net warming of the lower atmosphere and surface, it has become standard to include the effects of this stratospheric cooling in estimating radiative forcing due to carbon dioxide.

climate models (e.g., Schneider and Mass 1975). Today, two different numbers are in common use. Both are based on changes in global and annual mean-surface or near-surface temperature. *Equilibrium sensitivity* is defined as the long-term near-surface temperature increase after atmospheric carbon dioxide has been doubled from preindustrial levels but thereafter held constant until the Earth reaches a new steady state, as described in the preceding paragraph. *Transient climate response* or TCR is defined by assuming that carbon dioxide increases by 1% per year and then recording the temperature increase at the time carbon dioxide doubles (about 70 years after the increase begins). TCR depends on how quickly the climate adjusts to forcing, as well as on equilibrium sensitivity. The climate's adjustment time itself depends on equilibrium sensitivity and on the rate and depth to which heat is mixed into the ocean, because the depth of heat penetration tends to be greater in models with greater sensitivity (Hansen et al. 1985; Wigley and Schlesinger 1985). Accounting for ocean heat uptake complicates many attempts at estimating sensitivity from observations, as outlined below.

Equilibrium sensitivity depends on the strengths of feedback processes involving water vapor, clouds, and snow or ice extents (see, e.g., Hansen et al. 1984; Roe and Baker 2007). Small changes in the strengths of feedback processes can create large changes in sensitivity, making it difficult to tightly constrain climate sensitivity by restricting the strength of each relevant feedback process. As a result, research aimed at constraining climate sensitivity—and evaluating the sensitivities generated by models—is not limited to studies of these individual feedback processes. Studies of observed climate responses on short time scales (e.g., the response to volcanic eruptions or the 11-year solar cycle) and on long time scales (e.g., the climate of last glacial maximum 20,000 years ago) also play central roles in the continuing effort to constrain sensitivity. The quantitative value of each of these observational constraints is limited by the quality and length of relevant observational records, as well as the necessity in several cases to simultaneously restrict ocean heat uptake and equilibrium sensitivity. Equilibrium warming is directly relevant when considering paleocli-

mates, where observations represent periods that are very long compared to the climate's adjustment time. The transient climate response is more directly relevant to the attribution of recent warming and projections for the next century. For example, Stott et al. (2006) show that global mean warming due to well-mixed greenhouse gases over the 20th Century, in the set of models they consider, is closely proportional to the model's TCR. In the following, we discuss individual feedback processes as well as these additional observational constraints on sensitivity.

Equilibrium warming in an AOGCM is difficult to obtain because the deep ocean takes a great deal of time to respond to changes in climate forcing. To avoid unacceptably lengthy computer simulations, equilibrium warming usually is estimated from a modified climate model in which the ocean component is replaced by a simplified, fast-responding "slab ocean model." This procedure makes the assumption that horizontal redistribution of heat in the ocean does not change as the climate responds to the perturbation. Current climate models generate a range of equilibrium and transient climate sensitivities. For the models in the CMIP3 archive utilized in the Fourth Assessment of the IPCC, the range of equilibrium sensitivity is 2.1 to 4.4°C with a median of 3.2°C. This ensemble of models was not constructed to systematically span the plausible range of uncertainty in climate sensitivity; rather, each development team simply provided its best attempt at climate simulation. Complementary to this approach is one in which a single climate model is modified in a host of ways to explore more systematically the sensitivity variations associated with the range of uncertainty in various key parameters. Results with a Hadley Centre model give a 5 to 95 percentile range of ~2 to 6°C for equilibrium sensitivity (Piani et al. 2005; Knutti et al. 2006).

Charney (1979) provided a range of equilibrium sensitivities to CO_2 doubling of 1.5 to 4.5°C, based on the two model simulations available at the time. Evidently, the range of model-implied climate sensitivity has not contracted significantly over three decades. The current range, however, is based on a much larger number of models subjected to a far more comprehensive comparison to observations and containing more detailed treatments of clouds and other processes that are fundamental to climate sensitivity. We understand in much more detail why models differ in their equilibrium climate sensitivities: the source of much of this spread lies in differences in how clouds are modeled in AOGCMs. Questions remain as to whether or not the substantial spread among models is a good indication of the uncertainty in climate sensitivity, given all the constraints on this quantity of which we are aware. There also is a desire to know the prospects for constraining equilibrium climate sensitivity more sharply in the near future.

The variation among models is less for TCR than for equilibrium warming, a consequence of the interrelationship between the climate's adjustment time and its sensitivity to forcing noted above (Covey et al. 2003). The full range for TCR in the CMIP3 archive is 1.3 to 2.6°C, with a median of 1.6°C and 25 to 75% quartiles of 1.5 to 2.0°C (Randall et al. 2007). Systematic exploration of model input parameters in one Hadley Centre model gives a range of 1.5 to 2.6°C (Collins, M., et al. 2006).

The equilibrium and transient sensitivities in some models developed by U.S. centers contributing to CMIP3 are listed in Table 4.1. In the last column, the larger of the two GISS ModelE values is obtained using a full ocean model in which the circulation is allowed to adjust. All other values of equilibrium warming in the table are obtained with the ocean component replaced by a slab ocean model. The close agreement in transient climate sensitivity among models in this subset should not be overinterpreted, given the larger range among the full set of CMIP3 models.

Climate sensitivity is not a model input. It emerges from explicitly resolved physics, sub-grid-scale parameterizations, and numerical approximations used by the models—many of which differ from model to model—particularly those related to clouds and ocean mixing. The climate sensitivity of a model can be changed by modifying parameters that are poorly constrained by observations or theory. Influential early papers by Senior and Mitchell (1993, 1996) demonstrated how a seemingly minor

MODEL	TCR (°C)	Equilibrium Warming (°C)*
CSM1.4	1.4	2.0
CCSM2	1.1	2.3
CCSM3	1.5	2.5
GFDL CM2.0	1.6	2.9
GFDL CM2.1	1.5	3.4
GISS Model E		2.7 to 2.9

Table 4.1 Equilibrium and Transient Sensitivities in Some U.S. Models Contributing to CMIP3

*Equilibrium warming was assessed by joining a simplified slab ocean model to the atmosphere, land, and sea-ice AOGCM components.

[Sources of Information in table. First three lines – J.T. Kiehl et al. 2006: The climate sensitivity of the Community Climate System Model: CCSM3. *J. Climate*, **19**, 2584–2596. Next two lines – R.J. Stouffer et al. 2006: GFDL's CM2 global coupled climate models. Part IV: Idealized climate response. *J. Climate*, **19**, 723–740. Last line – J. Hansen et al. 2007: Climate simulations for 1880–2003 with GISS ModelE. *Climate Dynamics*, **29**(7–8), 661–696.]

modification to the cloud-prediction scheme alters climate sensitivity. In the standard version of the model, the effective size of cloud drops was fixed. In two other versions, this cloud-drop size was tied to the total amount of liquid-water cloud through two different empirical relationships. The equilibrium sensitivity ranged from 1.9 to 5.5°C in these three models. In general, the nonlinear dependence of equilibrium sensitivity on the strength of feedback processes allows relatively small changes in feedbacks to generate large changes in sensitivity (see, e.g., Hansen et al. 1984; Roe and Baker 2007).

Studies of the CCSM family of models provide another example of this problem. Kiehl et al. (2006) found that a variety of factors is responsible for differences in climate sensitivity among the models of this family. However, the lower TCR of CCSM2 (relative to CSM1.4 and CCSM3), evident in Table 4.1, results primarily from a single change in the model's algorithm for simulating convective clouds. Table 4.2 shows how equilibrium sensitivity varied during development of the most recent GFDL models. The dramatic drop in sensitivity between model versions p10 and p12.5.1 was unex-

**Table 4.2
Equilibrium Global
Mean Near-Surface
Warming Due to
Doubled
Atmospheric
Carbon Dioxide
from Intermediate
("p") Model
Versions Leading to
GFDL's CM2.0 and
CM2.1**

MODEL VERSION	Equilibrium Warming (°C)*
p7	3.87
p9	4.28
p10	4.58
p12.5.1	2.56
p12.7	2.65
p12.10b	2.87
p12b	2.83
CM 2.0	2.90
CM 2.1	3.43

*Equilibrium warming was assessed by joining a simplified slab ocean model to the atmosphere, land, and sea-ice AOGCM components.

[Source of information for table: Personal communication with Thomas Knutson, NOAA GFDL laboratory.]

pected. It followed a reformulation of the model's treatment of processes in the lower atmospheric boundary layer, which, in turn, affected how low-level clouds in the model respond to climate change.

4.1.2 Observational Constraints on Sensitivity

Climate models in isolation have not yet converged on a robust value of climate sensitivity. Furthermore, the actual climate sensitivity in nature might not be found in the models' range of sensitivities, since all the models may share common deficiencies. However, observations can be combined with models to constrain climate sensitivity. The observational constraints include the response to volcanic eruptions; aspects of the internal variability of climate that provide information on the strength of climatic "restoring forces"; the response to the 11-year cycle in solar irradiance; paleoclimatic information,

particularly from the peak of the last Ice Age some 20,000 years ago; aspects of the seasonal cycle; and, needless to say, the magnitude of observed warming over the past century.

4.1.2.1 VOLCANIC ERUPTIONS

Volcanoes provide a rapid change in radiative forcing due to the scattering and absorption of solar radiation by stratospheric volcanic aerosol. Of special importance, recovery time after the eruption contains information about climate sensitivity that is independent of uncertainties in the magnitude of the radiative forcing perturbation (e.g., Lindzen and Giannitsis 1998). Larger climate sensitivity implies weaker restoring forces on Earth's temperature, and, therefore, a slower relaxation back toward the unperturbed climate. However, this time scale also is affected by the pathways through which heat anomalies propagate into the ocean depths, with deeper penetration increasing the relaxation time. Several modeling studies have confirmed that this relaxation time after an eruption increases as climate sensitivity increases in GCMs when holding the ocean model fixed (Soden et al. 2002; Yokohata et al. 2005), encouraging the use of volcanic responses to constrain sensitivity. On the other hand, Boer, Stowasser, and Hamilton (2007) study two models with differing climate sensitivity and different ocean models; they highlight the difficulty in determining which model has the higher sensitivity from the surface-temperature responses to volcanic forcing in isolation, without quantitative information on ocean heat uptake.

Some studies have argued that observations of responses to volcanoes imply that models are overestimating climate sensitivity (e.g., Douglass and Knox 2005; Lindzen and Giannitsis 1998). These studies argue that observed relaxation times are shorter than those expected if climate sensitivity is as large as in typical AOGCMs. Studies that directly examine the volcanic responses in AOGCMs, however, find no such gross disagreement with observations (Wigley et al. 2005; Boer, Stowasser, and Hamilton 2007; Frame et al. 2005) consistent with earlier studies (e.g., Hansen et al. 1996; Santer et al. 2001). They nevertheless consistently suggest (Frame et al. 2005; Yokohata et al. 2005) that climate sensitivities as large as

6°C are inconsistent with observed relaxation times. It is important to note that these "observational" studies of climate sensitivity that do not utilize GCMs still make use of models, but they use simple energy balance "box" models rather than GCMs. The value of these studies depends on the relevance of the simple models as well as on the techniques for estimating parameters in models that control climate sensitivity. From these analyses, one can infer that further research isolating changes in ocean heat content after eruptions, such as that of Church, White, and Arblaster (2005), will be needed to strengthen constraints on climate sensitivity provided by responses to volcanic eruptions.

4.1.2.2 NATURAL CLIMATE VARIABILITY

Natural variability of climate also provides a way of estimating the strength of the restoring forces that determine climate sensitivity. Just as investigators learn something about sensitivity by watching the climate recover from a volcanic eruption, they can hope to obtain similar information by watching the climate relax from an unforced period of unusual global warmth or cold. This approach to constraining the response to a perturbation by examining the character of a system's natural variability, discussed by Leith (1975) in the context of climate sensitivity, is referred to as "fluctuation-dissipation" analysis in other branches of physics. In the case of equilibrium statistical mechanics, this relationship between characteristics of natural variability and response to an external force has been placed on a firm theoretical footing, but application to the climate is more heuristic, generally depending on approximation of the climate system by a linear stochastically forced model. The power of the approach is illustrated by Gritsun and Branstator (2007) in a study of the extratropical atmosphere's response to a perturbation in tropical heating. A recent attempt to apply this approach to climate sensitivity can be found in Schwartz (2007). This technique deserves more attention, with careful analysis of uncertainties. Its value likely will be determined by its ability to infer an AOGCM's sensitivity from an analysis of its internal variability.

4.1.2.3 SOLAR VARIATIONS

The 11-year solar cycle has potential for providing very useful information on climate sen-

sitivity. Total solar irradiance is known to vary by roughly 0.1% over this cycle (Frölich 2002). The expected response in global mean temperature is only ~0.1°C, so the technique is limited in value by the quality and length of the observational record, both of which restrict our ability to isolate this small signal. Recent results show promise in more cleanly identifying the climatic response to this cyclic perturbation (Camp and Tung 2007). Since ultraviolet wavelengths play a disproportionately larger role in these cyclic variations, detailed representations of the stratosphere and mesosphere, where ultraviolet radiation is absorbed, along with ozone chemistry are required for quantitative analysis of climatic response to the solar cycle (e.g., Shindell et al. 2006). Solar variations also have been invoked repeatedly to explain early 20th Century warming and to connect the Little Ice Age to the Maunder Minimum in sunspot number. While these connections may very well have a valid basis, using them to constrain climate sensitivity remains difficult as long as variations in insolation on time scales longer than the 11-yr cycle are not better quantified. To illustrate the difficulty, we note the substantial reduction in estimated insolation variations in the 20th Century between the Third and Fourth IPCC Assessments (Forster et al. 2007). Further analyses of responses to the sunspot cycle in models and observations seem likelier to lead to stronger constraints on climate sensitivity in the near term.

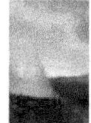

4.1.2.4 GLACIAL-INTERGLACIAL VARIATIONS

The glacial-interglacial fluctuations of the Pleistocene (the Ice Ages) are thought to be forced by changes in the Earth's orbit on time scales of 20,000 years and longer—the astronomical theory of the Ice Ages. Since this theory assumes that the mean temperature of the Earth can be altered by changing the distribution of the incoming solar flux without changing its global mean, it suggests important limitations to simple models based solely on global mean radiative forcing. For the limited purpose of constraining climate sensitivity, we need not understand how glacial-interglacial variations of ice sheets and of carbon dioxide are forced by changes in the Earth's orbit. Since we have knowledge from ice cores of greenhouse gas concentrations at the peak of the last major glacial advance 20,000 years ago as well as considerable information on the extent of continental ice sheets, one may ask if climate models can simulate the ocean-surface temperatures inferred from a variety of proxies, given these greenhouse gas concentrations and ice sheets (Manabe and Broccoli 1985). A logical assumption is that models that are more sensitive to doubling of carbon dioxide would also simulate larger cooling during the low carbon dioxide levels 20,000 years ago. Crucifix (2006) describes some of the difficulties with this simple picture. Annan and Hargreaves (2006) argue that the tropics and Antarctica are regions where this connection may be the strongest. Model results generated in the Paleoclimate Modelling Intercomparison Project (Braconnet et al. 2007a, b; Crucifix et al. 2006)) provide a valuable resource for analyzing these relationships. Despite these complications, several studies agree that past climates are difficult to reconcile with the low end of the equilibrium-sensitivity range generated by models (e.g., Hansen et al. 1993; Covey, Sloan, and Hoffert 1996). Models of the last glacial maximum also provide some of the strongest evidence that climate sensitivity is very unlikely to be larger than 6°C (Annan et al. 2005; Annan and Hargreaves 2006). As paleoclimatic reconstructions for this period improve, these simulations will become of greater quantitative value. Uncertainty in Ice Age aerosol concentrations may be the most difficult obstacle to overcome.

4.1.2.5 Seasonal Variation

The seasonal cycle is a familiar forced climate response to changes in the Earth-sun geometry and, therefore, should yield information on climate sensitivity. Although the seasonal cycles of global (Lindzen 1994) and hemispheric (Covey et al. 2000) mean temperature are not themselves strongly related to equilibrium climate sensitivity, regional variations and other aspects of the seasonal cycle may constrain sensitivity. Knutti et al. (2006) provide an example of a methodology using ensembles of climate model simulations to search for variables, or combinations of variables, that correlate with climate sensitivity (see also Shukla et al. 2006). If such a variable that predicts climate sensitiv-

ity in models is found, investigators can then examine its value in observations and hope thereby to constrain climate sensitivity. Knutti et al. (2006) use a neural network to look for aspects of the seasonal cycle with this predictive capability, with some success. Their study favors sensitivity in the middle of the typical model range (near 3°C).

The work of Qu and Hall (2006) provides an especially straightforward example of this approach. They do not address climate sensitivity directly but only the strength of one feedback mechanism that contributes to sensitivity: snow-albedo feedback (the decrease in reflection of solar radiation by snow as the snowcover retreats in a warming climate). Qu and Hall demonstrate that the strength of this feedback in models is strongly correlated to the seasonal cycle of the snow cover simulated by the models. Comparison of observed and simulated seasonal cycles of snow cover then suggest which model simulations of snow albedo feedback are the most reliable. These studies suggest that detailed comparisons of modeled and observed seasonal cycles should provide valuable information on climate sensitivity in the future.

The observed 20th Century warming is a fundamental constraint on climate models, but it is less useful than one might think in constraining sensitivity because of the large uncertainty in forcing due to anthropogenic aerosols. Twentieth Century simulations are important in demonstrating the consistency of certain combinations of sensitivity, aerosol forcing, and ocean-heat uptake, but they do not provide a sharp constraint on sensitivity in isolation (Kiehl 2007). Further discussion of 20th Century simulations can be found in Chapter 5.

Rather than focusing on one particular observational constraint or on models in isolation, attempts to combine some or all of these observational constraints with model simulations are recognized as the most productive approaches to constraining climate sensitivity (Bierbaum et al. 2003; Randall et al. 2007; Stott and Forest 2007). As an example, while model ensembles in which parameters are varied sys-

---

2 Estimating the probability of very high climate sensitivities above the high end of the CMIP3 model range, even if these probabilities are low, can be relevant for analyses of unlikely but potentially catastrophic climate change. It is not within the scope of this report to attempt to quantify these probabilities.

tematically can include models with sensitivities larger than 6°C (Stainforth et al. 2005; Roe and Baker 2007), these very high values can be excluded with high confidence through comparisons with observations of volcanic relaxation times and simulations of the last glacial maximum. As summarized by Randall et al. (2007) in the Fourth IPCC Assessment, these multiconstraint studies are broadly consistent with the spread of sensitivity in the CMIP3 models.[2]

4.2 FEEDBACKS

Better understanding of Earth's climate sensitivity, with potential reduction in its uncertainty, will require better understanding of a variety of climate feedback processes (Bony et al. 2006). We discuss some of these processes in more detail below.

4.2.1 Cloud Feedbacks

Clouds reflect solar radiation to space, cooling the Earth-atmosphere system. Clouds also trap infrared radiation, keeping the Earth warm. The integrated net effect of clouds on climate depends on their height, location, microphysical structure, and evolution through the seasonal and diurnal cycles. Cloud feedback refers to changes in cloud amounts and properties that can either amplify or moderate a climate change. Differences in cloud feedbacks in climate models have been identified repeatedly as the leading source of spread in model-derived estimates of climate sensitivity (beginning with Cess et al. 1990). The fidelity of cloud feedbacks in climate models therefore is important to the reliability of their prediction of future climate change.

Soden and Held (2006) evaluated cloud feedbacks in 12 CMIP3 AOGCMs and found weakly to strongly positive cloud feedback in the various models. The highest values of cloud feedback raise the equilibrium climate sensitivity (for CO_2 doubling) from values of about 2 K to roughly 4 K. In comparison with the earlier studies of Cess (1990) and Colman (2003), the spread of cloud feedbacks among GCMs has become somewhat smaller over the years but is still very substantial. Indeed, intermodel differences in cloud feedback are the primary reason

that models disagree in their estimates of equilibrium climate sensitivity; which (if any) models give accurate cloud simulations remains unclear (Randall et al. 2007) as debate over specific processes continues (Spencer et al. 2007)

Examples of competing hypotheses concerning high clouds (for which the infrared trapping effects are large) are the IRIS hypothesis of Lindzen, Chou, and Hou (2001) and the FAT (Fixed Anvil Temperature Hypothesis) of Hartmann and Larsson (2002). The IRIS hypothesis asserts that warmer temperatures cause the area coverage of clouds in the tropical upper troposphere to decrease, a negative feedback since these clouds are infrared absorbers. The FAT hypothesis asserts that the altitude of these tropical high clouds tends to increase with warming, minimizing the temperature change at the cloud tops—a positive feedback since the lack of warming at cloud top prevents the increase in outgoing radiation needed to balance the heat trapping of greenhouse gases. Observational studies aimed at evaluating these mechanisms are difficult because clouds in the tropics are strongly forced by circulations that are, in turn, driven by temperature gradients and not by the local temperature in isolation. These circulation effects must be eliminated to isolate effects relevant to global warming. Very high resolution simulations in localized regions have some potential to address these questions. The FAT hypothesis, in particular, has received some support from high-resolution modeling (Kuang and Hartmann 2007).

Although these studies focus on high clouds, the intermodel differences in model responses of low-level clouds are responsible for most of the spread of cloud feedback values in climate models (Bony et al. 2006). While tempting, assuming that this implies that low-cloud feedbacks are more uncertain than high-cloud feedbacks probably is premature. The strengths and weaknesses of cloud-cover simulations for present-day climate are described in Chapter 5.

As discussed in Chapter 6, a new class of much higher resolution global atmospheric simulations promises fundamental improvements in cloud simulation. Using the surrogate climate change framework of Cess (1990) in which ocean temperatures are warmed uniformly, Miura et al. (2005) carried out experiments using a global model with 7-km resolution, obtaining results suggestive of negative cloud feedback outside the tropics, and Wyant et al. (2006) describe results from a multigrid technique in which high-resolution cloud models are embedded in each grid box of a traditional GCM. Much work will be required with these new types of models before they can be given substantial weight in discussions of the most probable value for cloud feedback, but they suggest that real-world feedback is less positive than the typical CMIP3 AGCMs and that midlatitude cloud feedbacks may be more important than hitherto assumed. Results from this new generation of models will be of considerable interest in the coming years.

Several questions remain to be answered about cloud feedbacks in GCMs. Physical mechanisms underlying cloud feedbacks in different models must be better characterized. How best to judge the importance of model biases in simulations of current climate and in simulations of cloud changes in different modes of observed variability is not clear. In particular, how to translate these biases into levels of confidence in simulations of cloud feedback processes in climate change scenarios is unclear. New satellite products such as those from active radar and lidar systems should play a central role in cloud research in coming years by providing more comprehensive space-time cloud datasets.

4.2.2 Water-Vapor Feedbacks

Analysis of radiative feedbacks in the CMIP3 models (Soden and Held 2006) reaffirms that water-vapor feedback—the increase in heat trapping due to the increase in water vapor as the lower atmosphere warms—is fundamental to the models' climate sensitivity. The strength of their water-vapor feedback typically is close in magnitude to but slightly weaker than that obtained by assuming that relative humidity remains unchanged as the atmosphere warms.

A trend toward increasing column water vapor in the atmosphere consistent with model predictions has been documented from microwave satellite measurements (Trenberth, Fasullo, and Smith 2005), and excellent agreement for this

quantity has been found between satellite observations and climate models constrained by the observed ocean-surface temperatures (Soden 2000). These studies increase confidence in the models' vapor distributions more generally, but column water vapor is dominated by changes in the lower troposphere, whereas water-vapor feedback is strongest in the upper troposphere where most outgoing terrestrial radiation to space originates. The results of Soden and Held (2000) imply that at least half the global water-vapor feedback arises from the tropical upper troposphere in models in which relative humidity changes are small. Studies of vapor trends in this region are therefore of central importance. Soden et al. (2005) present analysis of radiance measurements, implying that relative humidity has remained unchanged in the upper tropical troposphere over the past few years, which, combined with temperature measurements, provides evidence that water vapor in this region is increasing.

Observations of interannual variability in water vapor can help to judge the quality of model simulations. Soden et al. (2002) concluded that a GCM appropriately simulates water-vapor variations in the tropical upper troposphere during cooling associated with the Pinatubo volcanic eruption. Minschwaner, Essler, and Sawaengphokhai (2006) compared the interannual variability of humidity measured in the highest altitudes of the tropical troposphere with CMIP3 20th Century simulations. Both models and observations show a small negative correlation between relative humidity and tropical temperatures, due in large part to lower relative humidity in warm El Niño years and higher relative humidity in cold La Niña years. However, there is a suggestion that the magnitude of this covariation is underestimated in most models. There also is a tendency for models with larger interannual variations in relative humidity to produce larger reductions in this region in response to global warming, suggesting that this deficiency in interannual variability might be relevant for climate sensitivity. (This is another example, analogous to the Qu and Hall (2006) analysis of snow feedback, in which the strength of a feedback in models is correlated with a more readily observed aspect of climatic variability.) In short, the study of Minschwaner, Essler, and Sawaengphokhai (2006) suggests that water-vapor feedback in the very highest levels of the tropical troposphere may be over-estimated in models, but it does not imply that a significant correction is needed to the overall magnitude of the feedback.

Positive water-vapor feedback, resulting from increases in vapor that keep the relative humidity from changing substantially as the climate warms, has been present in all GCMs since the first simulations of greenhouse gas–induced warming (Manabe and Wetherald 1975). It represents perhaps the single most robust aspect of global warming simulations. Despite the fact that the distribution of water vapor in the atmosphere is complex, we are aware of no observational or modeling evidence that casts doubt of any significance on this basic result, and we consider the increase in equilibrium sensitivity to roughly 2°C from this feedback to be a solid starting point from which the more uncertain cloud feedbacks then operate.

4.3 TWENTIETH CENTURY RADIATIVE FORCING

Radiative forcing is defined as a change that affects the Earth's radiation balance at the top of the tropopause between absorbed energy received in the form of solar energy and emitted infrared energy to space, typically expressed in terms of changes to the equilibrium preindustrial climate. Uncertainties in 20th Century radiative forcing limit the precision with which climate sensitivity can be inferred from observed temperature changes. In this section, we briefly discuss the extent to which models provide consistent and reliable estimates of radiative forcing over the 20th Century. Further information is provided by Forster et al. (2007).

Radiative forcing in models can be quantified in different ways, as outlined by Hansen et al. (2005). For example, the radiative forcing for the idealized case of CO_2 doubling can be computed by (1) holding all atmospheric and surface temperatures fixed, (2) allowing the stratospheric temperatures to adjust to the new CO_2 levels, (3) fixing surface temperatures over both land and ocean and allowing the atmosphere to equilibrate, or (4) fixing ocean temperatures only and allowing the atmosphere and land to equilibrate. Comparing model forcings

in the literature is complex because of differing calculations in different papers. An important objective for the climate modeling community is to improve the consistency of its reporting of radiative forcing in models.

4.3.1 Greenhouse Gases

Greenhouse gases like carbon dioxide and methane have atmospheric lifetimes that are long, compared to the time required for these gases to be thoroughly mixed throughout the atmosphere. Trends in concentration, consistent throughout the world, have been measured routinely since the International Geophysical Year in 1958. Measurements of gas bubbles trapped in ice cores give the concentration prior to that date (with less time resolution). Nevertheless, the associated radiative forcing varies somewhat among climate models because GCM radiative calculations must be computationally efficient, necessitating approximations that make them less accurate than the best laboratory spectroscopic data and radiation algorithms. Using changes in well-mixed greenhouse gases measured between 1860 and 2000, Collins et al. (2006b) compared the radiative forcing of climate models (including CCSM, GFDL, and GISS) with line-by-line (LBL) calculations in which fewer approximations are made. The median LBL forcing at the top of the model by greenhouse gases is 2.1 W/m², and the corresponding median among the climate models is higher by only 0.1 W/m². However, the standard deviation among model estimates is 0.30 W/m² (compared to 0.13 for the LBL calculations). Based on these most-recent comparisons with LBL computations, we can reasonably assume that radiative forcing due to carbon dioxide doubling in individual climate models may be in error by roughly 10%.

4.3.2 Other Forcings

While increases in the concentration of greenhouse gases provide the largest radiative forcing during the 20th Century, other smaller forcings must be considered to quantitatively model the observed change in surface air temperature. The burning of fossil fuels that release greenhouse gases into the atmosphere also produces an increase in atmospheric aerosols (small liquid droplets or solid particles that are temporarily suspended in the atmosphere). Aerosols cool the planet by reflecting sunlight back to space. In addition, among other forcings are changes in land use that alter the reflectivity of the Earth's surface, as well as variations in sunlight impinging on the Earth.

4.3.2.1 AEROSOLS

Aerosols have short lifetimes (on the order of a week) that prevent them from dispersing uniformly throughout the atmosphere, in contrast to well-mixed greenhouse gases. Consequently, aerosol concentrations have large spatial variations that depend on the size and location of sources as well as changing weather that disperses and transports the aerosol particles. Satellites can provide the global spatial coverage needed to observe these variations, but satellite instruments cannot distinguish between natural and anthropogenic contributions to total aerosol forcing. The anthropogenic component can be estimated using physical models of aerosol creation and dispersal constrained by available observations.

Satellites increasingly are used to provide observational estimates of the "direct effect" of aerosols on the scattering and absorption of radiation. These estimates range from -0.35 +/– 0.25 W/m^2 (Chung et al. 2005) to -0.5 +/– 0.33 W/m^2 (Yu et al. 2006) to -0.8 +/– 0.1 W/m^2 (Bellouin et al. 2005). The fact that two of these three estimates do not overlap suggests incomplete uncertainty analysis in these studies. In particular, each calculation must decide how to extract the anthropogenic fraction of aerosol. Global direct forcing by aerosols is estimated by the IPCC AR4 as -0.2 +/– 0.2 W/m^2, according to models, and -0.5 +/– 0.4 W/m^2, based upon satellite estimates and models. This central estimate is smaller in magnitude than the 2001 IPCC estimate of -0.9 +/– 0.5 W/m^2.

In addition to their direct radiative forcing, aerosols also act as cloud condensation nuclei. Through this and other mechanisms, they alter the radiative forcing of clouds (Twomey 1977; Albrecht 1989; Ackerman et al. 2004). Complex interactions among aerosols and cloud physics make this "aerosol indirect effect" very difficult to measure, and model estimates of it vary widely. This effect was generally omitted from

the IPCC AR4 models, although, among the U.S. CMIP3 models, it was included in GISS ModelE where increased cloud cover due to aerosols results in a 20th Century forcing of – 0.8 W/m^2 (Hansen et al. 2007).

4.3.2.2 VARIABILITY OF SOLAR IRRADIANCE AND VOLCANIC AEROSOLS

Other climate forcings include variability of solar irradiance and volcanic aerosols. Satellites provide the only direct measurements of these quantities at the top of the atmosphere. Satellite measurements of solar irradiance are available from the late 1970s and now span about 3 of the sun's 11-year magnetic or sunspot cycles. Extracting a long-term trend from this relatively brief record (Wilson et al. 2003) is difficult. Prior to the satellite era, solar variations are inferred using records of sunspot area and number and cosmic ray–generated isotopes in ice cores (Foukal et al. 2006), which are converted into irradiance variations using empirical relations The U.S. CMIP3 models all use the solar reconstruction by Lean, Beer, and Bradley (1995) with subsequent updates.

Volcanic aerosols prior to the satellite era are inferred from surface estimates of aerosol optical depth. The radiative calculation requires aerosol amount and particle size, which is inferred using empirical relationships with optical depth derived from recent eruptions. The GFDL and GISS models use updated versions of the Sato et al. (1993) eruption history, while the CCSM uses Ammann et al. (2003). As with solar variability, different reconstructions of volcanic forcing differ substantially (see, e.g., Lindzen and Giannitsis 1998). Land-use changes also are uncertain, and they can be of considerable significance locally. Global models, however, typically show very modest global responses, as discussed in Hegerl et al. (2007).

Studies attributing 20th Century global warming to various natural and human-induced forcing changes clearly are hindered by these uncertainties in radiative forcing, especially in the solar and aerosol components. The trend in total solar irradiance during the last few decades (averaging over the sun's 11-year cycle) apparently is negative and thus cannot explain recent global warming (Lockwood and Fröhlich 2007). The connection between solar energy output changes and the warming earlier in the 20th Century is more uncertain. With the solar reconstructions assumed in the CMIP3 models, much of the early 20th Century warming is driven by solar variations, but uncertainties in these reconstructions do not allow confident attribution statements concerning this early-century warming. The large uncertainties in aerosol forcing are a more important reason that the observed late 20th Century warming cannot be used to provide a sharp constraint on climate sensitivity. We do not have good estimates of the fraction of greenhouse gas forcing that has been offset by aerosols.

4.4 OCEAN HEAT UPTAKE AND CLIMATE SENSITIVITY

As noted above, the rate of heat uptake by the ocean is a primary factor determining transient climate response (TCR): the larger the heat uptake by the oceans, the smaller the initial response of Earth's surface temperature to radiative forcing (e.g., Sun and Hansen 2003). Studies show (e.g., Völker, Wallace, and Wolf-Gladrow 2002) that CO_2 uptake by the ocean also is linked to certain factors that control heat uptake, albeit not in a simple fashion. In an AOGCM, the ocean component's ability to take up heat depends on vertical mixing of heat and salt and how the model transports heat between low latitudes (where heat is taken up by the ocean) and high latitudes (where heat is given up by the ocean). The models make use of several subgrid-scale parameterizations (see Chapter 2), which have their own uncertainties. Thus, as part of understanding a model's climate-sensitivity value, we must assess its ability to represent the ocean's mixing processes and the transport of its heat, as well as feedbacks among the ocean, ice, and atmosphere.

The reasons for differing model estimates of ocean uptake are incompletely understood. Assessments typically compare runs of the same model or output from different AOGCMs. Raper, Gregory, and Stouffer (2002) examined climate sensitivity and ocean heat uptake in a suite of then-current AOGCMs. They calculated the ratio of the change in heat flux (from the surface to the deep ocean) to the change in temperature (Gregory and Mitchell 1997) and

found in general that models with lower ocean-uptake efficiency had lower climate sensitivity, as expected (Hansen et al. 1985; Wigley and Schlesinger 1985). Uptake efficiency can be thought of as the amount of heat the ocean absorbs through mixing relative to the change in surface temperature (e.g., to reproduce the observed 20th Century warming despite a high climate sensitivity, a model needs large heat export to the deep ocean). Comparing the current generation of AOGCMs with the previous generation, however, Kiehl et al. (2006) found that the atmospheric component of the models is the primary reason for different transient climate sensitivities, and the ocean component's ability to uptake heat is of secondary importance. Ocean heat-uptake efficiency values calculated in this study differ substantially from those in Raper et al. (2002).

Despite these complexities, modern ocean GCMs are able to transport both heat (AchutaRao et al. 2006) and passive tracers such as chlorofluorocarbons and radiocarbon (Gent et al. 2006; Dutay et al. 2002) consistent with the limited observations available for these quantities. Better observations in the future—particularly of the enhanced ocean warming expected from the anthropogenic greenhouse effect—should provide stronger constraints on modeled ocean transports.

4.5 IMPACT OF CLIMATE SENSITIVITY ON USING MODEL PROJECTIONS OF FUTURE CLIMATES

This chapter has emphasized the global and annual mean of surface temperature change even though practical applications of climate change science involve particular seasons and locations. The underlying assumption is that local climate impacts scale with changes in global mean surface temperature (Santer et al. 1990). In that case, time histories of global mean temperature—obtained from a simple model of global mean temperature, run under a variety of forcing scenarios—could be combined with a single AOGCM-produced map of climate change normalized to the global mean surface temperature change. In that way, the regional changes expected for many different climate-forcing scenarios could be obtained from just one AOGCM simulation using one idealized forcing scenario such as atmospheric CO_2 doubling (Oglesby and Saltzman 1992) or 1% per year increasing CO_2 (Mitchell et al. 1999). This "pattern scaling" assumption also permits the gauging of effects on regional climate change that arise from different estimates of global climate sensitivity. For example, if an AOGCM with TCR = 1.5 K predicts temperature and precipitation changes ΔT and ΔP as a function of season and location in a 21st Century climate simulation, and if investigators believe that TCR = 1.0 K is a better estimate of the real world's climate sensitivity, then, under the pattern-scaling assumption, they would reduce the local ΔT and ΔP values by 50%.

Although it introduces its own uncertainties, the pattern-scaling assumption increasingly is used in climate impact assessments (e.g., Mitchell 2003; Ruosteenoja, Tuomenvirta, and Jylha 2007). For example, the annual mean temperature change averaged over the central United States during the 21st Century for any of the projections in the IPCC *Special Report on Emissions Scenarios* shows that about 75% of the variance among the CMIP3 models is explained by their differing global mean warming (B. Wyman, personal communication). (The central United States is defined in this context following Table 11.1 in Christensen et al. 2007.) Precipitation patterns, in contrast, do not scale as well as temperature patterns due to sharp variations between locally decreasing and locally increasing precipitation in conjunction with global warming.

Model Simulation of Major Climate Features

Although a typical use of atmosphere-ocean general circulation model (AOGCM) output for climate impact assessment focuses on one particular region such as a river basin or one of the 50 United States, knowing model simulation overall accuracy on continental to global scales is important. Fidelity in simulating climate on the largest scales is a necessary condition for credible predictions of future climate on smaller scales. Model developers devote great effort to assessing the level of agreement between simulated and observed large-scale climate, both for the present day and for the two centuries since the Industrial Era began. Unlike physical theories of such fundamentally simple systems as the hydrogen atom, AOGCMs cannot promise precise accuracy for every simulated variable on all relevant space and time scales. Nevertheless, before applying a model to a practical question, users should demand reasonable overall agreement with observations, with the definition of "reasonable" in part subjective and dependent on the problem at hand. Here we provide an overview of how well modern AOGCMs satisfy this criterion.

5.1 MEAN SURFACE TEMPERATURE AND PRECIPITATION

Simulations of monthly near-surface air temperature and precipitation provide a standard starting point for model evaluation since these fields are central to many applications. The two fields also illustrate the difficulty in designing appropriate metrics for measuring model quality.

By most measures, modern AOGCMs simulate the basic structure of monthly mean near-surface temperatures quite well. The globally averaged annual mean value generally lies within the observed range (~286 to 287 K) of modern and preindustrial values; this agreement, however, is in part a consequence of the "final tun-

ing" of the models' energy balance as described in Chapter 2 and by itself is not a stringent test of model quality. More relevant is consideration of space and time variations about the global annual mean (including the seasonal cycle). The overall correlation pattern between simulations and observations typically is 95 to 98%, and variation magnitudes typically agree within ±25% (Covey et al. 2003). This level of success has been retained in the latest generation of models that allow ocean and atmosphere to exchange heat and water without artificial adjustments (Randall et al. 2007). Nevertheless, as shown below, local errors in surface temperature that are clearly outside the bounds of observational uncertainty persist in the latest generation of models.

AOGCM simulations are considerably less accurate for monthly mean precipitation than for temperature. The space-time correlation between models and observations typically is only about 50 to 60% (Covey et al. 2003). As we discuss below, these poor correlations originate mainly in the tropics, where precipitation varies greatly over relatively small ranges of latitude and longitude. Strong horizontal gradients in the field lead to a significant drop in correlations with observations, even with only slight shifts in the modeled precipitation distribution. These modest correlations are relevant for precipitation at a particular location, but AOGCMs generally reproduce the observed broad patterns of precipitation amount and year-to-year variability (see Fig. 5.1 and Dai 2006). One prominent error is that models without flux adjustment typically fail to simulate the observed northwest-to-southeast orientation of a large region of particularly heavy cloudiness and precipitation in the southwest Pacific Ocean. Instead, these models tend to rotate this convergence zone into an east-west orientation, producing an unrealistic pair of distinct, parallel convection bands straddling the equator instead of a continuous Inter-Tropical Convergence Zone (ITCZ). The double-ITCZ error has been frustratingly persistent in climate models despite much effort to correct it.

Another discrepancy between models and observations appears in the average day-night cycle of precipitation. While the model's diurnal temperature cycle exhibits general agreement with observations, simulated cloud formation and precipitation tend to start too early in the day. Also, when precipitation is sorted into light, moderate, and heavy categories, models reproduce the observed extent of moderate precipitation (10 to 20 mm/day) but underestimate that of heavy precipitation and overestimate the extent of light precipitation (Dai 2006). Additional model errors appear when precipitation is studied in detail for particular regions [e.g., within the United States (Ruiz-Barradas and Nigam 2006)].

For illustration, we show examples from two of the U.S. models discussed in Chapter 4. In Fig. 5.1 (Delworth et al. 2006) and Fig. 5.2 (Collins et al. 2006a), simulated and observed maps of surface temperature and even precipitation appear rather similar at first glance. Constructing simulated-minus-observed difference maps, however, reveals monthly and seasonal mean temperature and precipitation errors up to 10°C and 7 mm/day, respectively, at some points. CCSM3 temperature-difference maps exhibit the largest errors in the Arctic (note scale change in Fig. 5.2d), where continental wintertime near-surface temperature is overestimated. AOGCMs find this quantity particularly difficult to simulate because, for land areas near the poles in winter, models must resolve a strong temperature inversion above the surface (warm air overlying cold air). For precipitation, GFDL difference maps reveal significant widespread errors in the tropics, most notably in the ITCZ region discussed above and in the Amazon River basin, where precipitation is underestimated by several millimeters per day. Similar precipitation errors appear in CCSM3 results (e.g., a 28% underestimate of Amazon annual mean). AOGCM precipitation errors have serious implications for Earth system models with interactive vegetation, because such models use simulated precipitation to calculate plant growth (see Chapter 6). Errors of this magnitude would produce an unrealistic distribution of vegetation in an Earth system model, for example, by spuriously deforesting the Amazon basin.

In summary, modern AOGCMs generally simulate continental and larger-scale mean surface temperature and precipitation with considerable accuracy, but the models often are not reliable for smaller regions, particularly for precipitation.

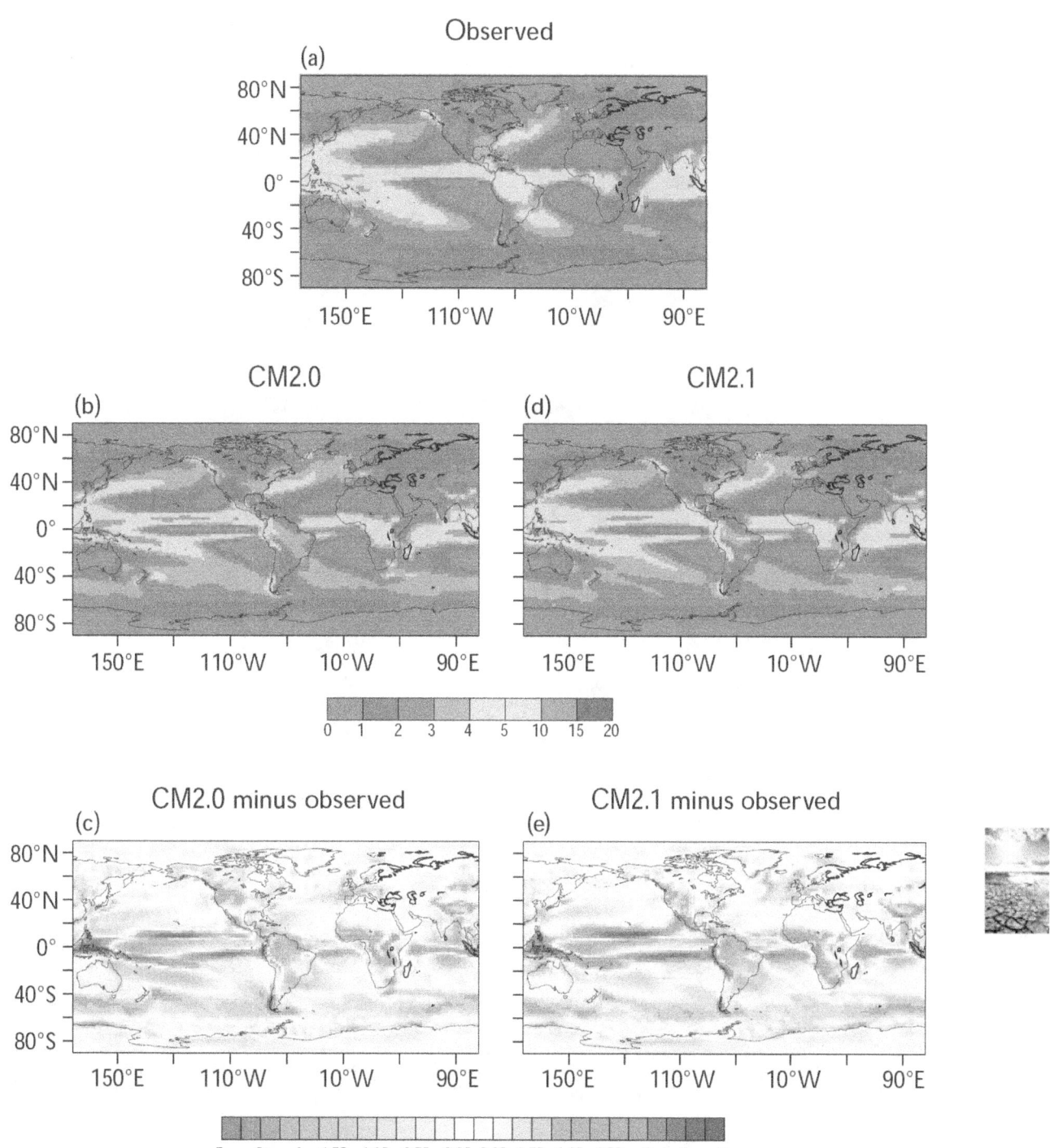

Figure 5.1a–e. Observed and GFDL Model-Simulated Precipitation (mm/day).

Observed image from P. Xie and P.A. Arkin 1997: Global precipitation: A 17-year monthly analysis based on gauge observations, satellite estimates, and numerical model outputs. *Bulletin American Meteorological Society,* **78**, 2539–2558. [Other images from Fig. 17 in T.L. Delworth et al. 2006: GFDL's CM2 global coupled climate models. Part I: Formulation and simulation characteristics. *J. Climate,* **19**, 643–674. Images reproduced with permission of the American Meteorological Society.]

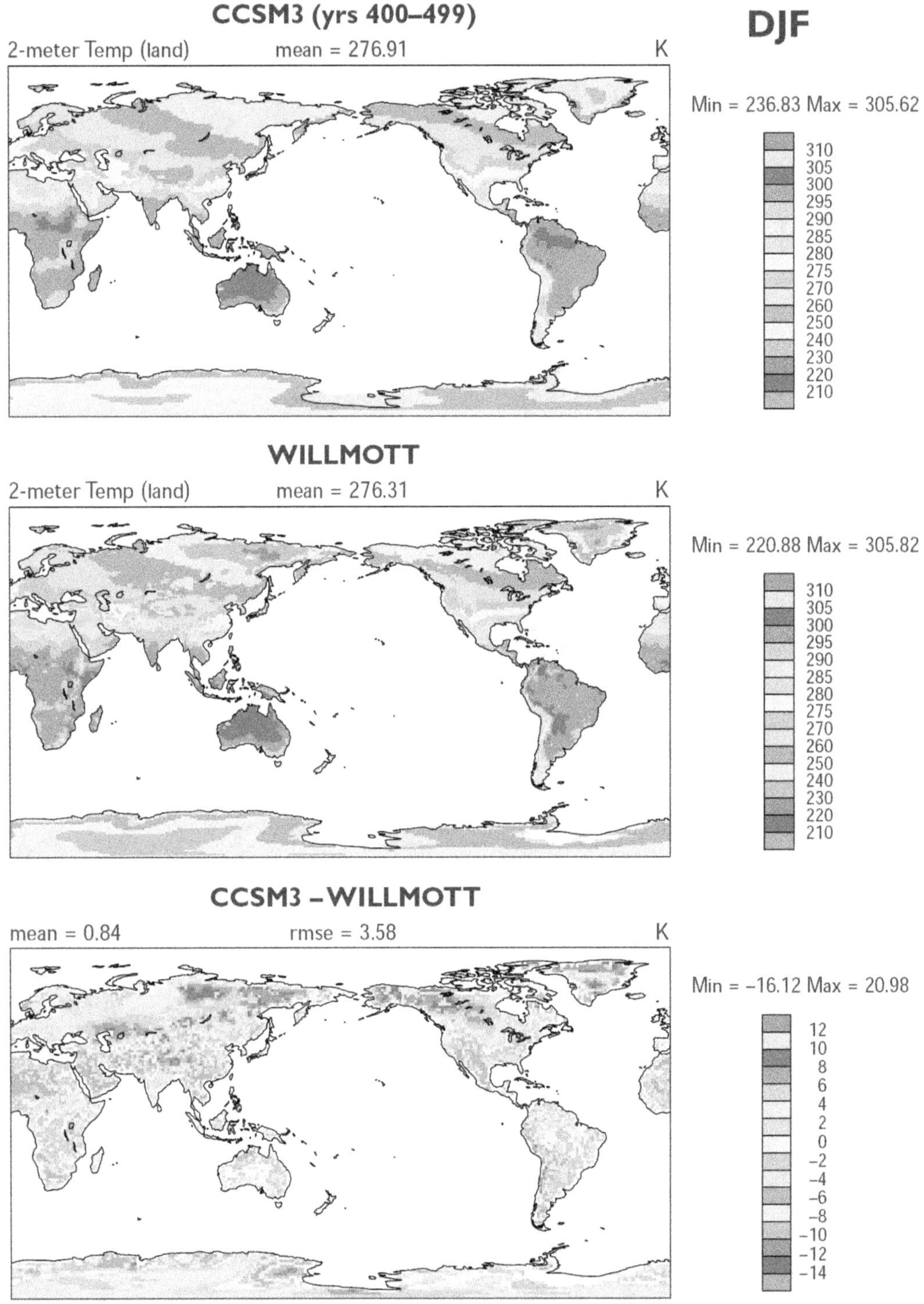

CCSM3 (yrs 400–499)

DJF

2-meter Temp (land)   mean = 276.91   K

Min = 236.83 Max = 305.62

310
305
300
295
290
285
280
275
270
260
250
240
230
220
210

WILLMOTT

2-meter Temp (land)   mean = 276.31   K

Min = 220.88 Max = 305.82

310
305
300
295
290
285
280
275
270
260
250
240
230
220
210

CCSM3 –WILLMOTT

mean = 0.84   rmse = 3.58   K

Min = –16.12 Max = 20.98

12
10
8
6
4
2
0
–2
–4
–6
–8
–10
–12
–14

Figure 5.2a–c. CCSM3 December-January-February Simulated (top panel), Observed (middle panel), and Simulated-Minus-Observed (bottom panel) Near-Surface Air Temperature for Land Areas (°C).

Note change in scale from 5.2a to 5.2c. [Figures from W. Collins et al. 2006: The Community Climate System Model Version 3 (CCSM3). *J. Climate*, **19**(11), 2122–2143. Reproduced with permission of the American Meteorological Society.]

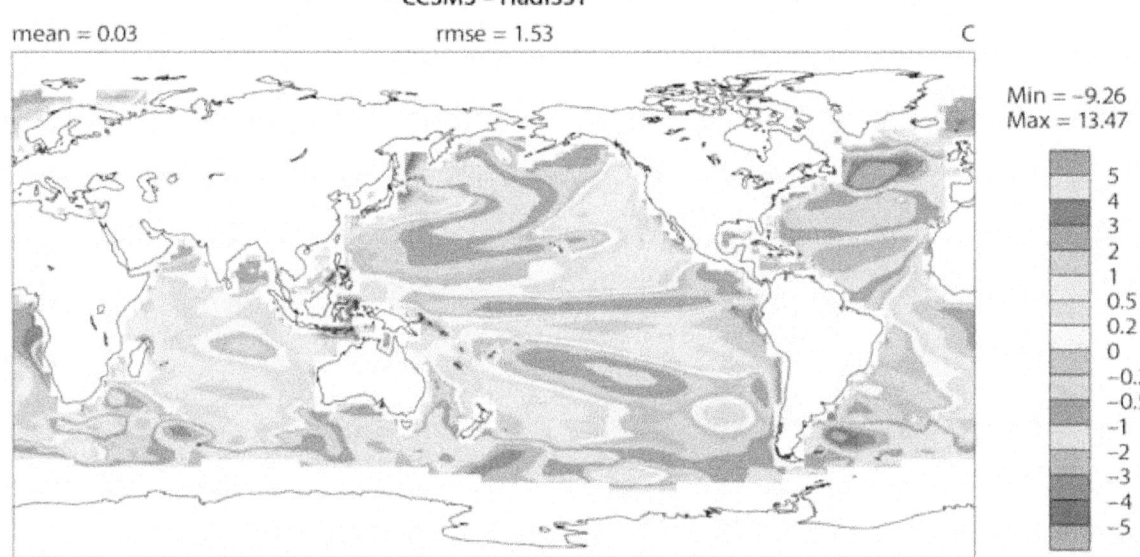

CCSM3 – HadISST

mean = 0.03   rmse = 1.53   C

Min = −9.26
Max = 13.47

5
4
3
2
1
0.5
0.2
0
−0.2
−0.5
−1
−2
−3
−4
−5

Figure 5.2d. CCSM3 Annual Mean Simulated-Minus-Observed Sea Surface Temperature (°C).

[Figure from W. Collins et al. 2006: The Community Climate System Model Version 3 (CCSM3). *J. Climate*, **19**(11), 2122–2143. Reproduced with permission of the American Meteorological Society.]

5.2 TWENTIETH CENTURY TRENDS

Modern AOGCMs are able to simulate not only the time-average climate but also changes (trends) in climate over the past 140 years. For example, Fig. 5.3 shows results from the three U.S. models and the "average" CMIP3 model. Plotted in the figure are curves of globally averaged annual mean near-surface temperature from model simulations and the observational value as determined from the U.K. Climatic Research Unit (CRU) gridded observational database. Two curves are plotted for the CMIP3 models. The first shows the average over all CMIP3 models, and the second, the average over only CMIP3 models that included the effects of volcanic eruptions. Results from individual U.S. models are shown for separate ensemble members (dotted lines) and for the average over all ensemble members (continuous lines). Individual members of a particular model ensemble differ from each other because they were run from different initial conditions. Precise initial conditions, especially deep-ocean

temperature and salinity, are not known for 1860. The spread among individual simulations from the same model (the dotted-line curves) thus indicates uncertainty in model-simulated temperature arising from lack of knowledge about initial conditions.

These results demonstrate that modern climate models exhibit agreement with observed global mean near-surface temperature trends to within observational uncertainty, despite imprecise initial conditions and uncertain climate forcing and heat uptake by the deep ocean (Min and Hense 2006). Models achieve this agreement only if they include anthropogenic emissions of greenhouse gases and aerosols. No plausible combination of natural climate-forcing factors allows models to explain the global warming observed over the last several decades. Indirect solar effects [e.g., involving cosmic rays and clouds (Svensmark 2007)] are not generally included in AOGCM simulations. These effects have been proposed occasionally as causes of global warming, although over the past 20 years their trends would, if anything, lead to cooling

(Lockwood and Fröhlich 2007). Unless the models grossly underestimate the climate system's natural internally generated variability or are all missing a large unknown forcing agent, the conclusion is that most recent warming is anthropogenic (IPCC 2007b).

Nevertheless, total climate forcing during the 20th Century is not accurately known, especially the aerosol component (see Chapter 2). Aerosol forcing used in these simulations, however, is derived from aerosol parameterizations constrained by satellite and ground-based measurements of the aerosols themselves and was not designed to obtain a fit to observed global mean temperature trends. The observed trend in surface temperature can result from models with different aerosol forcing (Schwartz 2007). Thus, 20th Century temperature records cannot distinguish models that would warm by differing amounts for the same total forcing.

Note that climate sensitivity is not prescribed in AOGCMs. Instead, this sensitivity emerges as a result of a variety of lower-level modeling choices. In contrast to simple energy-balance models that predict only the global mean temperature using a limited representation of climate physics, an AOGCM's climate sensitivity is difficult to specify a priori. More fundamentally, AOGCMs, unlike simpler climate models, have far fewer adjustable parameters than the number of observations available for model evaluation (Randall et al. 2007). Thus, an AOGCM's multidimensional output can be compared to observations independent of this adjustment (e.g., using observed trends in regional temperature). Agreement between modeled and observed trends has been described for temperature trends on each inhabited continent (Min and Hense 2007); for trends in climate extremes, such as heat-wave frequency and frost-day occurrence (Tebaldi et al. 2006); and for trends in surface pressure and Arctic sea ice (see Chapter 9 in IPCC 2007), all of which complement comparisons between modeled and observed time-averaged climate discussed in the following sections.

Figure 5.3a. Simulation of 20th Century Globally Averaged Surface Temperature from GFDL CM2.1.

"CRU" is the value based on the Climate Research Unit gridded observational dataset, "IPCC Mean" is the average value of all CMIP3 models, and "IPCC Mean Volc" is the average of all CMIP3 models that included volcanic forcing. Individual realizations of the CMIP3 20th Century experiment are denoted by the dotted curves labeled "run(1–3)," and the ensemble mean is marked "Mean."

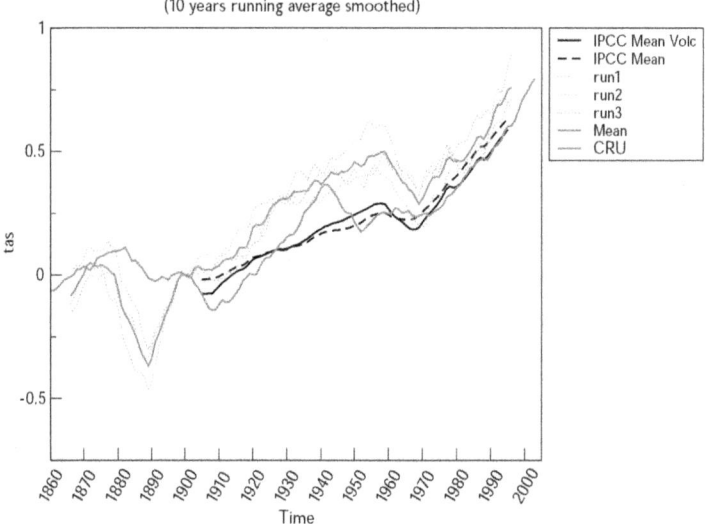

Global Warming relative to 1900 for giss_model_e_r

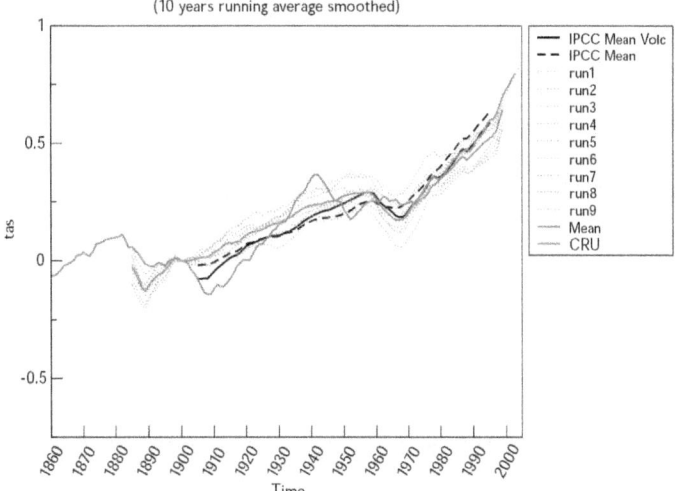

Figure 5.3b. Simulation of 20th Century Globally Averaged Surface Temperature from GISS Model E-R.

Curve labels are the same as in Fig. 5.3a.

Global Warming relative to 1900 for ncar_ccsm3_0

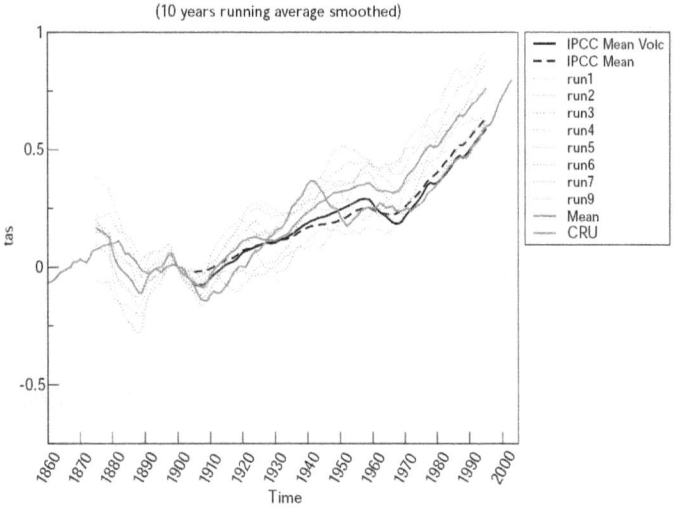

Figure 5.3c. Simulation of 20th Century Globally Averaged Surface Temperature from CCSM3.

Curve labels are the same as in Fig. 5.3a.

Global Warming Relative to 1900 for American Models

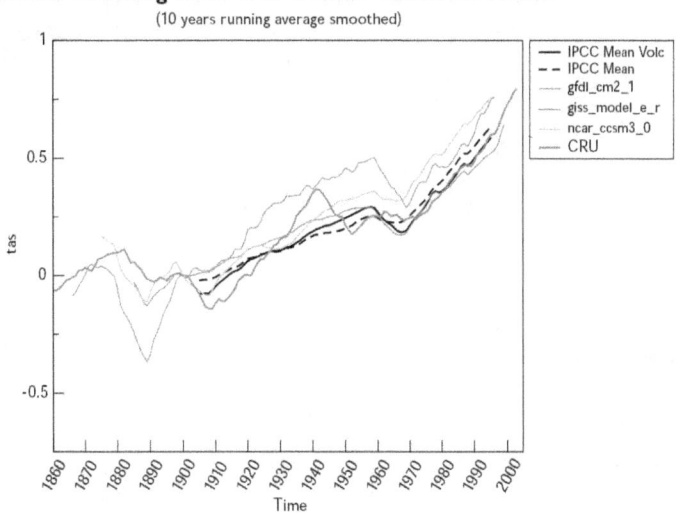

Figure 5.3d. Comparison of Simulations of 20th Century Globally Averaged Surface Temperature from the Three U.S. CMIP3 Models.

Model curves represent ensemble means for CCSM3 (ncar_ccsm3_0), GISS Model E-R (giss_e_r), and GFDL CM2.1 (gfdl_cm2_1). "CRU," "IPCC Mean," and "IPCC Mean Volc" labels are the same as in Fig. 5.3a.

As an example of 20th Century temperature trends on continental-to-global spatial scales and multidecadal time scales, Fig. 5.4 shows global maps for different time periods between 1880 and 2003 as observed and simulated by GISS ModelE (Hansen et al. 2006; also see Knutson et al. 2006). The figure shows general agreement between model and observations not only for the overall period but also for segments 1880 to 1940 and 1979 to 2003, which encompass periods of early and late 20th Century warming. For 1940 to 1979, the model simulates only a small change in global mean temperature in agreement with observations, but it fails to simulate the strong north polar cooling observed for this period. As a result, the model-simulated global mean-temperature change (upper right corner of each frame) is slightly positive rather than slightly negative as observed. Part of this discrepancy may result from chaotic fluctuations within observed climate that the model cannot synchronize correctly due to inprecise knowledge of the initial conditions in the 19th Century period. These chaotic fluctuations generally are more important in re-

gional trends than in the global average, where uncorrelated fluctuations in different regions tend to cancel. For both 20th Century warming periods, the model simulates, but underestimates, the high-latitude amplification of global warming. Additional discrepancies between AOGCMs and observations appear at smaller scales. For example, model-simulated trends do not consistently match the observed lack of 20th Century warming in the central United States (Kunkel et al. 2006).

5.2.1 Trends in Vertical Temperature

While models simulate the 20th Century warming observed at the surface, agreement is less obvious with tropospheric observations from satellites and weather balloons. This issue was the focus of CCSP SAP 1.1 (CCSP 2006). Since 1979 (beginning of the satellite record), globally averaged warming in the troposphere according to climate models is within the range of available observations. Within the tropics, the model-simulated troposphere warms more rapidly than observed (see CCSP 2006, Fig. 5.4 F–

Surface Temperature Change Based on Local Linear Trends (°C)

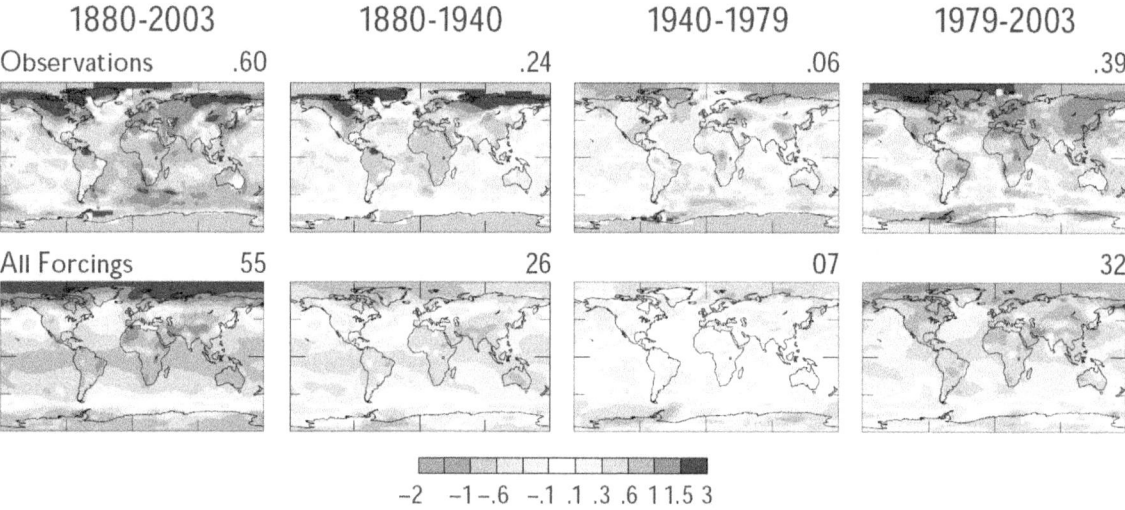

Figure 5.4. Near-Surface Temperature Changes as Observed (top panels) and as Simulated by GISS ModelE (bottom panels) for Selected Time Periods Between 1880 and 2003.

Numbers above upper right panel corners are global means. [Images from Fig. 9 in J. Hansen et al. 2007: Climate simulations for 1880–2003 with GISS ModelE. *Climate Dynamics*, **29**(7–8), 661–696. Reproduced with kind permission of Springer Science and Business Media.]

G). SAP 1.1 noted, however, that "Large structural uncertainties in the observations . . . make it difficult to reach more definitive conclusions regarding the significance and importance of model-data discrepancies" (CCSP 2006, p. 112 and Section 5.4).

Research since publication of SAP1.1 has continued to highlight uncertainties implicit in measuring the difference between surface and lower-atmospheric warming. For example, Thorne et al. (2007) found that the tropical atmosphere-to-surface warming ratio in both observations and model simulations is sensitive to the time period analyzed. Meanwhile, debate continues over the best way to process data from satellites (Christy et al. 2007) and weather balloons (Christy and Spencer 2005). AOGCMs continue to differ from most published observations on the ratio of atmosphere-to-surface warming in the tropics since the beginning of satellite observations (e.g., as shown by Thorne et al. 2007, Fig. 3), with the ratio being larger in the models than is seen in decadal observational trends.

Paradoxically, trends are more consistent between models and observations on interannual time scales. AOGCM simulation of tropical atmospheric warming involves mainly subgrid-scale parameterizations. As discussed in Chapter 2, these are not as trustworthy as explicitly computed processes, but internal variability [primarily due to El Niño–Southern Oscillation (ENSO)] provides a useful test of the models' ability to redistribute heat realistically. AOGCMs simulate very well the portion of tropical temperature trends due to interannual variability (Santer et al. 2005). In addition, explaining how atmospheric water vapor increases coincidentally with surface temperature is difficult (Trenberth, Fasullo, and Smith 2005; Santer et al. 2007; Wentz et al. 2007) unless lower tropospheric temperature also increases coincidentally with surface temperature. While deficiencies in model subgrid-scale parameterizations are certainly possible, trends in poorly documented forcing agents (see Chapter 4) may prove important in explaining the discrepancy over the longer time scales. Future research is required to resolve the issue because tropospheric observations at face value suggest a trend toward greater tropical instability, which has im-

plications for many aspects of model projections in the tropics.

5.2.2 Model Simulation of Observed Climate Variability

The following sections discuss a number of specific climate phenomena directly or indirectly related to near-surface temperature, precipitation, and sea level. Numerous studies of climate change have focused on one or two of these phenomena, so a great deal of information (and occasional debate) has accumulated for each of them. Here we attempt to summarize the points that would best give users of AOGCM model output a general sense of model reliability or unreliability. Although the following sections individually note different types of climate variation, the reader should recognize that the total amount of natural climate variability forms background "noise" that must be correctly assessed to identify the "signal" of anthropogenic climate change. Natural variability in turn separates into an externally forced part (e.g., from solar energy output and volcanic eruptions) and internally generated variability just as weather varies on shorter time scales because of the system's intrinsic chaotic character. As noted above, long-term trends in both solar and volcanic forcing during the past few decades have had a cooling rather than warming effect. It follows that if global warming during this period is not anthropogenic, then the climate system's internal variation is the most likely alternative explanation.

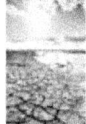

Control runs of AOGCMs (in which no changes in external climate forcing are included) provide estimates of the level of internally generated climate variability. Control runs generally obtain realistic near-surface temperature variability on annual-to-decadal time scales, although they typically underestimate variability in areas of the Pacific and Indian Ocean where ENSO and the Pacific Decadal Oscillation (PDO) (see below) predominate (Stouffer, Hegerl, and Tett 2000). Unfortunately, the longest time periods that are directly relevant to separating natural from anthropogenic climate change are the least observed. Assessing variations of surface temperature for time periods longer than 50 to 100 years depends on paleodata such as ice-core composition and tree-ring thickness. Interpretation of these data is made difficult by sparse geographical coverage and also is complicated by natural variations in external climate forcing.

5.2.2.1 EXTRA-TROPICAL STORMS

Climate models have developed from numerical weather-prediction models whose performance has been judged primarily on their ability to forecast midlatitude weather. The success of forecast models in their simulation of midlatitude cyclones and anticyclones has resulted in continuous growth in the value of numerical weather prediction. The ability of GCMs to generate realistic statistics of midlatitude weather also has been central in climate model development. This is true not only because midlatitude weather is important in its own right, but also because these storms are the primary mechanism by which heat, momentum, and water vapor are transported by the atmosphere, making their simulation crucial for simulation of global climate. Indeed, a defining feature of atmospheric general circulation models (AGCMs) is that they compute midlatitude eddy statistics and associated eddy fluxes through explicit computation of the life cycles of individual weather systems and not through some turbulence or parameterization theory. Computing the evolution of individual eddies may seem very inefficient when primary interest is in long-term eddy statistics, but the community clearly has judged for decades that explicit eddy simulation in climate models is far superior to attempts to develop closure theories for eddy statistics. The latter theories typically form the basis for Earth system models of intermediate complexity (EMICs), which are far more efficient computationally than GCMs but provide less convincing simulations.

Two figures illustrate the quality of simulated midlatitude eddy statistics from coupled AOGCMs used in IPCC AR4. Shown for the GFDL CM2.1 in Fig. 5.5a is wintertime variance of the north-south velocity component at 300 hPa (in the upper troposphere). This quantity represents the magnitude of variability in the upper troposphere associated with day-to-day weather. In Fig. 5.5b, the wintertime poleward eddy heat flux or covariance between temperature and north-south velocity is shown at 850 mb (in the lower troposphere). For these calculations, the monthly means were sub-

tracted before computing variances. In each case, eddy statistics are compared to estimates of observed statistics obtained from NCEP/NCAR Reanalysis (B.Wyman, personal communication). When analyzing eddy statistics, the data are typically filtered to retain only those time scales, roughly 2 to 10 days, associated with midlatitude weather systems. The two quantities chosen here, however, are sufficiently dominated by these time scales that they are relatively insensitive to the monthly filtering used here. In winter, Northern Hemisphere storms are organized into two major oceanic storm tracks over the Pacific and Atlantic oceans. Historically, atmospheric models of horizontal resolutions of 200 to 300 km typically are capable of simulating midlatitude storm tracks with realism comparable to that shown in the figure. Eddy amplitudes often are a bit weak and often displaced slightly equatorward. In spectral models with resolution coarser than 200 to 300 km, simulation of midlatitude storm tracks typically deteriorates significantly (see, e.g., Boyle 1993). General improvements in most models in the CMIP3 database over previous generations of models, as described in Chapter 1, are thought to be partly related to the fact that most of these models now have grid sizes of 100 to 300 km or smaller. Although even-finer resolution results in better simulations of midlatitude-storm structure, including that of warm and cold fronts and interactions among these storms and coastlines and mountain ranges, improvements in midlatitude climate on large scales tend to be less dramatic and systematic. Other factors besides horizontal resolution are considered important for details of storm track structure. Such factors include distribution of tropical rainfall, which is sensitive to parameterization schemes used for moist convection, and interactions between stratosphere and troposphere, which are sensitive to vertical resolution. Roeckner et al. (2006), for example, illustrate the importance of vertical resolution for midlatitude circulation and storm track simulation.

Lucarini et al. (2006) provide a more detailed look at the ability of CMIP3 models to simulate the space-time spectra of observed eddy statistics. These authors view the deficiencies noted, which vary in detail from model to model, as serious limitations to model credibility. As indicated in Chapter 1, however, our ability is lim-

ited in translating measures of model biases into useful measures of model credibility for 21st Century projections, and the implications of these biases in eddy space-time spectra are not self-evident. Indeed, in the context of simulating eddy characteristics generated in complex turbulent flows in the laboratory (e.g., Pitsch 2006), the quality of atmospheric simulations, based closely on fluid dynamical first principles, probably should be thought of as one of the most impressive characteristics of current models. As an example of a significant model deficiency that plausibly can be linked to limi-

tations in climate projection credibility, note that the Atlantic storm track, as indicated by the maximum velocity variance in Fig. 5.5a, follows a latitude circle too closely and the observed storm track has more of a southwest-northeast tilt. This particular deficiency is common in CMIP3 models (van Ulden and van Oldenborgh 2006) and is related to difficulty in simulating the blocking phenomenon in the North Atlantic with correct frequency and amplitude. Van Ulden and van Oldenborgh make the case that this bias is significant for the quality of regional climate projections over Europe.

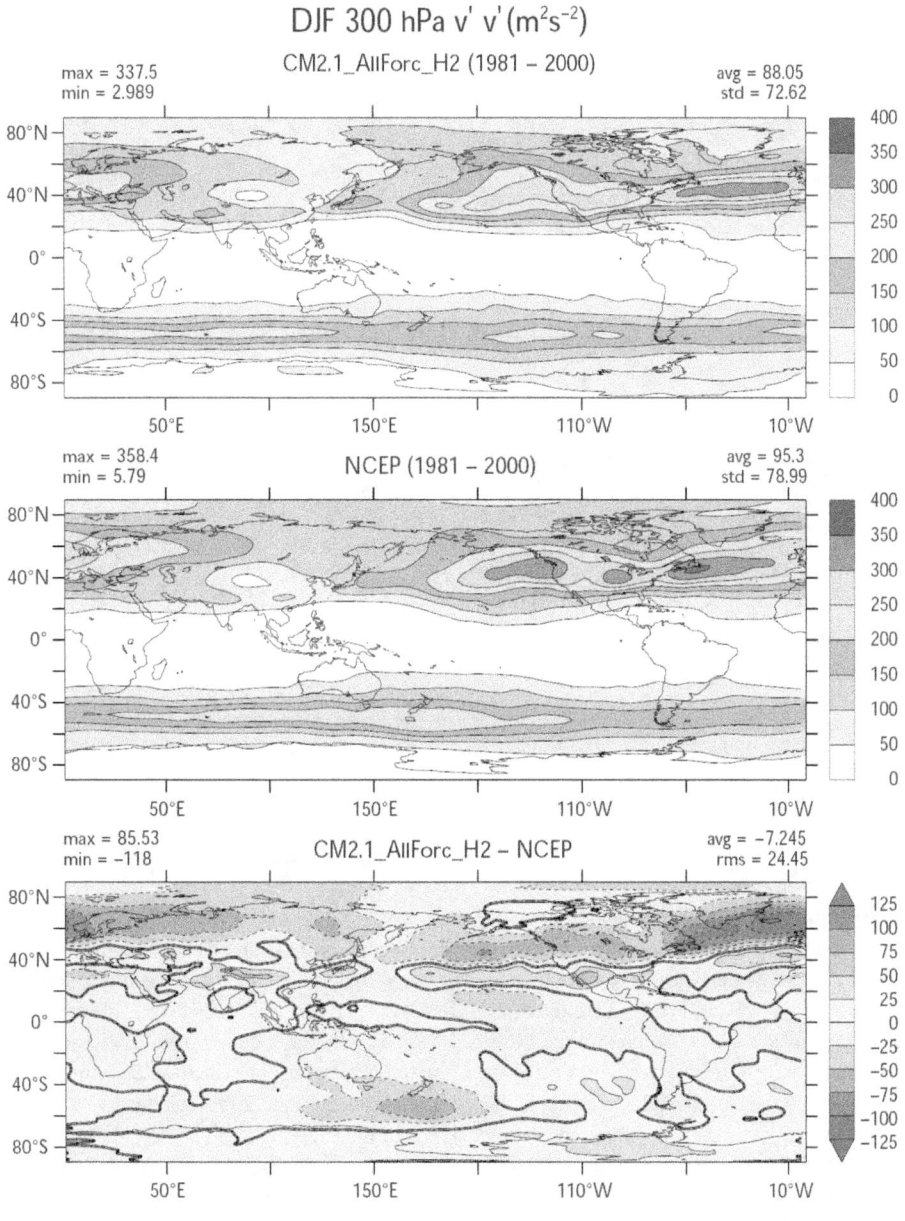

Figure 5.5a. Top: Variance of North-South Velocity at 300 hPa as Simulated by GFDL CM2.1 Model in Years 1981 to 2000 of One Realization of 20C3M Simulation, as Contributed to the CMIP3 Database.

Units are m²/s². Middle: Same quantity as obtained from NCEP/NCAR Reanalysis (Kalnay et al. 1996). Bottom: Model minus observations.

Figure 5.5b. Top: Covariance of North-South Velocity and Temperature at 850 hPa as Simulated by GFDL CM2.1 Model in Years 1981 to 2000 of One Realization of 20C3M Simulation, as Contributed to the CMIP3 Database.

Units are K-m/s. Middle: Same quantity as obtained from NCEP/NCAR Reanalysis (Kalnay et al. 1996). Bottom: Model minus observations.

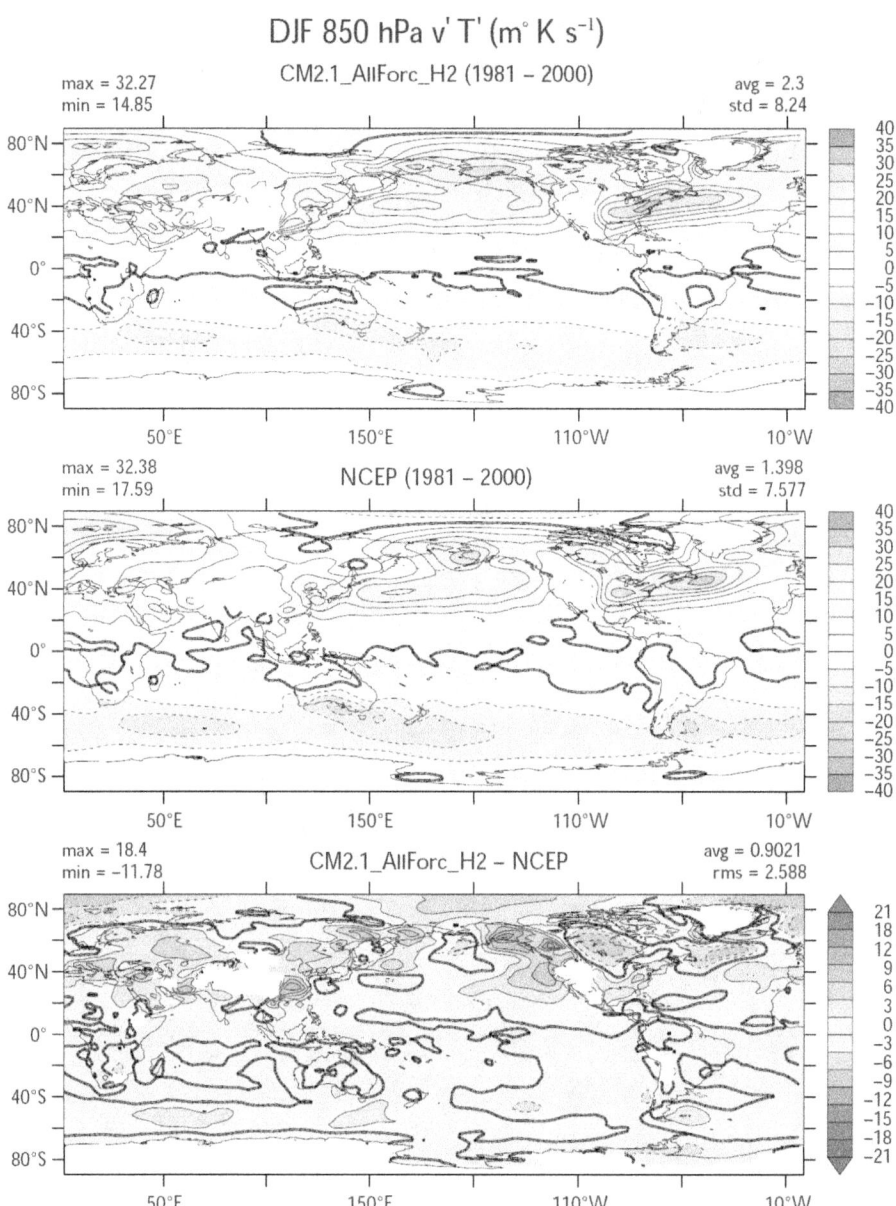

5.2.2.2 TROPICAL STORMS

Tropical storms (hurricanes in the Atlantic and typhoons in the Pacific and Indian oceans) are too small to be simulated reliably in the class of global-climate models currently used for climate projections. There is hope for simulating regional climate aspects that control the genesis of tropical depressions, however. Vitart and Anderson (2001), for example, identified tropical storm-like vortices in simulations with models of this type, demonstrating some skill

in simulating the effects of El Niño on Atlantic storm frequency.

Simulations with atmospheric models are steadily moving to higher resolutions (e.g., Bengtsson, Hodges, and Esch 2007). The recent 20-km–resolution simulation with an atmospheric model over prescribed ocean temperatures by Oouchi et al. (2006) is indicative of the kinds of modeling that will be brought to bear on this problem in the next few years. Experience with tropical storm forecasting suggests that this resolution should be adequate for describing many aspects of the evolution of mature tropical storms and possibly the generation of storms from incipient disturbances, but probably not tropical storm intensity. A promising alternative approach is described by Knutson et al. (2007), in which a regional model of comparable resolution (18 km) is used in a downscaling framework (see Chapter 3) to simulate the Atlantic hurricane season. Given observed year-to-year variations in the large-scale atmosphere structure over the Atlantic Ocean, the model is capable of simulating year-to-year variations in hurricane frequency over a 30-year period with a correlation of 0.7 to 0.8. It also captures the observed trend toward greater hurricane frequency in the Atlantic during this period. These results suggest that downscaling using models of this resolution may be able to provide a convincing capability for tropical storm frequency projections into the future, although these projections still will rely on the quality of global model projections for changes in sea-surface temperature, atmospheric stability, and vertical shear.

5.2.2.3 Monsoons

A monsoonal circulation is distinguished by its seasonal reversal after the sun crosses the equator into the new summer hemisphere. Rain is most plentiful in, if not entirely restricted to, summer within monsoonal climates, when continental rainfall is supplied mainly by evaporation from the nearby ocean. This limits the reach of monsoon rains to the distance over which moisture can be transported onshore (Privé and Plumb 2007). Variations in the monsoon's spatial extent from year to year determine which inland regions experience drought.

Over a billion people are dependent on the arrival of monsoon rains for water and irrigation for agriculture. The Asian monsoon during boreal summer is the most prominent example of a monsoon circulation dominating global rainfall during this season. However, the summer rainfall maximum and seasonal reversal of winds also indicate monsoon circulations in West Africa and the Amazon basin. In addition, during boreal summer, air flows off the eastern Pacific Ocean toward Mexico and the American Southwest while, over the Great Plains, mois-

ture from the Gulf of Mexico brings an annual peak in rainfall. Thus, the climate in these regions also is described as monsoonal.

Because of the Asian monsoon's geographical extent, measures of the fidelity of Asian monsoonal simulations can differ depending on specific regional focus and the metrics being used. Kripalani et al. (2007) judged that 3/4 of the 18 analyzed coupled models match the timing and magnitude of the summertime peak in precipitation over East Asia between 100 and 145°E and 20 to 40°N evident in the NOAA-NCEP Climate Prediction Center's Merged Analysis of Precipitation (CMAP, Xie, and Arkin 1997). However, only half of these models were able to reproduce the gross observed spatial distribution of monsoon rainfall and its migration along the coast of China toward the Korean peninsula and Japan. Considering a broader range of longitude (40 to 180°E) that includes the Indian subcontinent, Annamalai, Hamilton, and Sperber (2007) found that 6 of 18 AOGCMs significantly correlated with the observed spatial pattern of CMAP precipitation from June through September. (These six models also produced relatively realistic simulation of ENSO variability, which is known to influence interannual variations in the Asian summer monsoon.) Kitoh and Uchiyama (2006) computed the spatial correlation and root-mean-square error of simulated precipitation over a similar region and found, for example, the GFDL models in the top tercile with a spatial correlation exceeding 0.8.

During boreal winter, Asian surface winds are directed offshore: from the northeast over India and the northwest over East Asia. Hori and Ueda (2006) provide correlations between observed spatial distributions of surface pressures and 850-mb zonal winds during the East Asian winter monsoon with winds and pressures simulated by nine CMIP3 models. Correlations for zonal winds, for example, vary from 0.96 to 0.75. Monsoonal simulations in these models clearly vary considerably in quality, more so perhaps than other circulation features. Observed year-to-year variability of the West African monsoon is related to remote ocean temperatures in the North and South Atlantic and Indian oceans (Rowell et al. 1992; Zhang and Delworth 2006) as well as to temperatures in the nearby Gulf of Guinea. Cook and Vizy (2006) found that slightly more than half of 18 analyzed coupled models reproduced the observed precipitation maximum over land from June through August. Of these models, only six (including GISS ModelE-H and both GFDL models) reproduced the observed anticorrelation between Gulf of Guinea ocean temperature and Sahel rainfall.

The late 20th Century Sahel drought was a dramatic change in the Earth's hydrological cycle that plausibly must be simulated by climate models if we are to have any confidence in their ability to project future climate in this region. Atmospheric models, when run over observed oceanic temperatures, simulate this drought reasonably well (Hoerling et al. 2006). In these models, the drought is at least partly forced by warming of the Northern Hemisphere oceans, particularly the North Atlantic, with respect to Southern Hemisphere oceans, especially the Indian Ocean and Gulf of Guinea. Although the consensus is that these variations in ocean temperature gradients are at least partly due to natural variability, they may have been partly anthropogenically forced. Analysis of CMIP3 simulations of the 20th Century by Biasutti and Giannini (2006), supporting the earlier modeling study of Rotstayn and Lohmann (2002), suggests that aerosol forcing in these models played a part in generating this drought by cooling the North Atlantic with respect to other ocean basins. A small number of coupled models simulate droughts of the observed magnitude, including GFDL models (Held and Soden 2006), but why some models are more realistic in this regard than others is not understood.

Rainfall over the Sahel and Amazon are anticorrelated: when the Gulf of Guinea warms, rainfall generally is reduced over the Sahel but increases over South America. Amazon rainfall also depends on the eastern equatorial Pacific, and, during an El Niño, rainfall is reduced in the Nordeste region of the Amazon. Li et al. (2006) compare the hydrological cycle of 11 CGCMs over the Amazon during the late 20th and 21st centuries. Based on a comparison to CMAP rainfall, the GISS ModelE-R is among the best.

The ability of climate models to simulate Northern Hemisphere summer rainfall over the U.S.

Great Plains and Mexico was summarized by Ruiz-Barradas and Nigam (2006). Models generally have more difficulty in simulating summer rainfall in the Great Plains than winter rainfall, and this disparity probably should be thought of as reflecting the quality of future rainfall projections as well. Strengths and weaknesses vary considerably across the models. As an example, GISS ModelE-H closely matches the annual precipitation cycle over the Great Plains and Mexico and is one of two models to simulate interannual precipitation variations significantly correlated with observed variability during the second half of the 20th Century.

Initial monsoon evaluations simulated by the most recent generation of climate models have emphasized the seasonal time scale. However, subseasonal variations, such as break periods when the monsoon rains are interrupted temporarily, are crucial to forecasting the monsoon's impact on water supply. Simulating the diurnal cycle and the local hour of rainfall also is important to partitioning rainfall between runoff and transpiration, and these are important topics for future model evaluation. Transports of moisture by regional circulations beneath model resolution (such as low-level jets along the Rockies and Andes and tropical cyclones) contribute to the onshore transport of moisture. In general, models show some success at simulating gross seasonal features of various monsoon circulations, but studies are limited on variations of the smaller spatial and time scales important to specific watersheds and hydrological projections.

5.2.2.4 MADDEN-JULIAN OSCILLATIONS

The Madden-Julian Oscillation (MJO) consists of large-scale eastward-propagating patterns in humidity, temperature, and atmospheric circulation that strengthen and weaken tropical rainfall as they propagate around the Earth in roughly 30 to 60 days. This pattern often dominates tropical precipitation variability on time scales longer than a few days and less than a season, creating such phenomena as 1- to 2-week breaks in Asian monsoonal rainfall and weeks with enhanced hurricane activity in the eastern North Pacific and the Gulf of Mexico. Inadequate prediction of the evolution of these propagating structures is considered a main impediment to more useful extended-range

weather forecasts in the tropics, and improved simulation of this phenomenon is considered an important metric for the credibility of climate models in the tropics.

Nearly all models capture the pattern's essential feature, with large-scale eastward propagation and with roughly the correct vertical structure. But propagation often is too rapid and amplitudes too weak. Recent surveys of model performance indicate that simulations of MJO remain inadequate. For example, Lin et al. (2006), in a study of many CMIP3 models, conclude that "… current GCMs still have significant problems and display a wide range of skill in simulating the tropical intraseasonal variability," while Zhang et al. (2005) in another multimodel comparison study, state that "… commendable progress has been made in MJO simulations in the past decade, but the models still suffer from severe deficiencies …." As an example of recent work, Boyle et al. (2008) attempted, with limited success, to determine whether two U.S. CMIP3 models could maintain a preexisting strong MJO pattern when initialized with observations [from the Tropical Ocean Global Atmosphere–Coupled Ocean Atmosphere Response Experiment (called TOGA-COARE) field experiment].

The difficulty in simulating MJO is related to the phenomenon's multiscale nature: the propagating pattern itself is large enough to be resolvable by climate models, but the convection and rainfall modulated by this pattern, which feed back on the large-scale environment, occur on much smaller, unresolved scales. In addition to this dependence on parameterization of tropical convection, a long list of other effects has been shown by models and observational studies to be important for MJO. These effects include the pattern of evaporation generated as MJO propagates through convecting regions, feedback from cloud-radiative interactions, intraseasonal ocean temperature changes, the diurnal cycle of convection over the ocean, and the vertical structure of latent heating , especially the proportion of shallow cumulus congestus clouds and deep convective cores in different phases of oscillation (Lin et al. 2004)].

A picture seems to be emerging that simulation difficulty may not be due to a single model de-

ficiency but is a result of the phenomenon's complexity, given the long list of factors thought to be significant. In several multimodel studies such as Lin et al. (2006), a few models do perform well. However, without a clearer understanding of how these factors combine to generate the observed characteristics of MJO, maintaining a good simulation when the model is modified for other reasons is difficult, as is applying the understanding gained from one model's successful simulation to other models. Whether models with superior MJO simulations should be given extra weight in multimodel studies of tropical climate change is unclear.

5.2.2.5 El Niño–Southern Oscillation

By the mid-20th Century, scientists recognized that a local anomaly. in rainfall and oceanic upwelling near the coast of Peru was in fact part of a disruption to atmospheric and ocean circulations across the entire Pacific basin. During El Niño, atmospheric mass migrates west of the dateline as part of the Southern Oscillation, reducing surface pressure and drawing rainfall into the central and eastern Pacific (Rasmussen and Wallace 1983). Together, El Niño and the Southern Oscillation, abbreviated in combination as ENSO, are the largest source of tropical variability observed during recent decades. Because of the Earth's rotation, easterly winds along the equator cool the surface by raising cold water from below, which offsets heating by sunlight absorption (e.g., Clement et al. 1996). Cold water is especially close to the surface in the east Pacific, while warm water extends deeper in the west Pacific so upwelling has little effect on surface temperature there. The westward increase in temperature along the equator is associated with a decrease in atmospheric pressure, reinforcing the easterly trade winds. El Niño occurs when easterly trade winds slacken, reducing upwelling and warming the ocean surface in the central and east Pacific.

Changes along the equatorial Pacific have been linked to global disruptions of climate (Ropelewski and Halpert 1987). During an El Niño event, the Asian monsoon typically is weakened, along with rainfall over eastern Africa, while precipitation increases over the American Southwest. El Niño raises the surface temperature as far poleward as Canada, while changes in the north Pacific Ocean are linked to decadal variations in ENSO (Trenberth and Hurrell 1994). In many regions far from eastern equatorial Pacific, accurate projections of climate change in the 21ˢᵗ Century depend upon the accurate projection of changes to El Niño. Moreover, the demonstration that ENSO alters climate across the globe indicates that even changes to the *time-averaged* equatorial Pacific during the 21ˢᵗ Century will influence climate far beyond the tropical ocean. For example, long-term warming of the eastern equatorial Pacific relative to the surrounding ocean will favor a weaker Asian monsoon year after year, even in the absence of changes to the size and frequency of El Niño events.

In general, coupled models developed for CMIP3 are far more realistic than those of a decade ago, when ENSO variability was comparatively weak and some models lapsed into permanent El Niño states (Neelin et al. 1992). Even compared to models assessed more recently by the El Niño Simulation Intercomparison Project (called ENSIP) and CMIP2 (Latif et al. 2001; AchutaRao and Sperber 2002), ENSO variability of ocean surface temperature is more realistic in CMIP3 simulations, although sea-level pressure and precipitation anomalies show little recent improvement (AchutaRao and Sperber 2006). Part of this progress is the result of increased resolution of equatorial ocean circulation that has accompanied increases in computing speed. Table 5.1 shows horizontal and vertical resolution near

the equator in oceanic components of the seven American coupled models whose output was submitted to CMIP3.

Along the equator, oceanic waves that adjust the equatorial temperature and currents to changes in the wind are confined tightly to within a few degrees of latitude. To simulate this adjustment, the ocean state is calculated at points as closely spaced as 0.27 degrees of latitude in the NCAR CCSM3. NCAR PCM has a half-degree resolution, while both GFDL models have equatorial resolution of a third of a degree. This degree of detail is a substantial improvement compared to previous generations of models. In contrast, the GISS AOM and ModelE-R calculate equatorial temperatures at grid points separated by four degrees of latitude. This is broad compared to the latitudinal extent of cold temperatures observed within the eastern Pacific. The cooling effect of upwelling is spread over a larger area, so the amplitude of the resulting surface temperature fluctuation is weakened. In fact, both the GISS AOM models and ModelE-R have unrealistic ENSO variations that are much smaller than observed (Hansen et al. 2007). This minimizes the influence of their simulated El Niño and La Niña events on climate outside the equatorial Pacific, and we will not discuss these two models further in this section.

In comparison to previous generations of global models, where ENSO variability was typically weak (Neelin et al. 1992), the AR4 coupled models generally simulate El Niño near the observed amplitude or even above (AchutaRao and Sperber 2006). The latter study compared sea-surface temperature (SST) variability within the tropical Pacific, calculated under preindustrial conditions. Despite its comparatively low two-degree latitudinal grid spacing, the GISS ModelE-H (among American models) most closely matches observed SST variability since the mid-19ᵗʰ Century, according to the HadISST v1.1 dataset (Rayner et al. 2003). The NCAR PCM also exhibits El Niño warming close to the observed magnitude. This comparison is based on spatial averages within three longitudinal bands, and GISS ModelE-H, along with NCAR models, exhibits its largest variability in the eastern band as observed. However, GISS ModelE-H underestimates variability since 1950, when the NCAR CCSM3 is closest to observa-

Table 5.1. Spacing of Grid Points at the Equator in the American Coupled Models Developed for AR4*

MODEL	Longitude	Latitude	Vertical Levels
GFDL CM2.0	1	1/3	50
GFDL CM2.1	1	1/3	50
GISS AOM	5	4	13
GISS ModelE-H	2	2	16
GISS ModelE-R	5	4	13
NCAR CCSM3	1.125	0.27	27
NCAR PCM	0.94	0.5	32

*Except for GISS models, spacing of grid points generally increases away from the equator outside the ENSO domain, so resolution is highest at the equator.

tions (Joseph and Nigam 2006). Although the fidelity of each model's ENSO variability depends on the specific dataset and period of comparison (c.f. Capotondi, Wittenberg, and Masina 2006; Merryfield 2006; van Oldenborgh, Philip, and Collins 2005), the general consensus is that GISS ModelE-H, both NCAR models, and GFDL CM2.0 have roughly the correct amplitude, while variability is too large by roughly one-third in GFDL CM2.1. Most models (including GISS ModelE-H and both NCAR models but excluding GFDL models) exhibit the largest variability in the eastern band of longitude, but none of the CMIP3 models matches the observed variability at the South American coast where El Niño was identified originally (AchutaRao and Sperber 2006; Capotondi, Wittenberg, and Masina 2006). This possibly is because the longitudinal spacing of model grids is too large to resolve coastal upwelling and its interruption during El Niño (Philander and Pacanowski 1981). Biases in atmospheric models (e.g., underestimating persistent stratus cloud decks along the coast) also may contribute (Mechoso et al. 1995).

El Niño occurs every few years, albeit irregularly. The spectrum of anomalous ocean temperature shows a broad peak between 2 and 7 years, and multidecadal variations occur in event frequency and amplitude. Almost all AR4 models have spectral peaks within this range of time scales. Interannual power is distributed broadly within the American models, as observed, with the exception of NCAR CCSM3, which exhibits strong biennial oscillations (Guilyardi 2006).

Although models generally simulate the observed magnitude and frequency of events, reproducing their seasonality is more elusive. Anomalous warming typically peaks late in the calendar year, as originally noted by South American fisherman. Among American models, this seasonal dependence is simulated only by NCAR CCSM3 (Joseph and Nigam 2006). Warming in GFDL CM2.1 and GISS ModelE-H is nearly uniform throughout the year, while warming in NCAR PCM is largest in December but exhibits a secondary peak in early summer. The mean seasonal cycle along the equatorial Pacific also remains a challenge for the models. Each year, the east Pacific cold tongue is observed to warm during boreal spring and cool again late in the calendar year. GFDL CM2.1 and NCAR PCM1 have the weakest seasonal cycle among American models, while GISS ModelE-H, GFDL 2.0, and NCAR CCSM3 are closest to the observed amplitude (Guilyardi 2006). Among the worldwide suite of CMIP3 models, amplitude of the seasonal cycle of equatorial ocean temperature generally varies inversely with the ENSO cycle's strength.

Several studies have compared mechanisms generating ENSO variability in CMIP3 models to those inferred from observations (e.g., van Oldenborgh, Philip, and Collins 2005; Guilyardi 2006; Merryfield 2006; Capotondi, Wittenberg, and Masina 2006). Models must simulate the change in ocean upwelling driven by changes in surface winds, which in turn are driven by regional contrasts in ocean temperature. In general, GFDL2.1 is ranked consistently among American models as providing the most realistic simulation of El Niño. This is not based primarily on its surface-temperature variability (which is slightly too large) but on its faithful simulation of the observed relationship between ocean temperature and surface wind, along with wind-driven ocean response. While SST variability in CMIP3 models is controlled by anomalies of either upwelling rate or temperature, these processes alternate in importance over several decades within GFDL CM2.1 as observed (Guilyardi 2006). Since the 1970s the upwelling temperature, rather than the rate, has been the predominant driver of SST variability (Wang 1995). A confident prediction of future El Niño amplitude requires both the upwelling rate and temperature, along with their relative amplitude, to be simulated correctly. This remains a challenge.

El Niño events are related to climate anomalies throughout the globe. Models with more realistic ENSO variability generally exhibit an anti-correlation with the strength of the Asian summer monsoon (e.g., Annamalai, Hamilton, and Spencer 2007), while 21st Century changes to Amazon rainfall have been shown to depend on projected trends in the tropical Pacific (Li et al. 2006). El Niño has a long-established relation to North American climate (Horel and Wallace 1981), assessed in CMIP3 models by

Joseph and Nigam (2006). This relation is strongest during boreal winter, when tropical anomalies are largest. Anomalous circulations driven by rainfall over the warming equatorial Central Pacific radiate atmospheric disturbances into midlatitudes amplified within the north Pacific storm track (Sardeshmukh and Hoskins 1988; Held, Lyons, and Nigam 1989; Trenberth et al. 1998). To simulate ENSO's influence on North America, models must represent realistic rainfall anomalies in the correct season so the connection is amplified by wintertime storm tracks. The connection between equatorial Pacific and North American climate is simulated most accurately by the NCAR PCM model (Joseph and Nigam 2006). In GFDL CM2.1, North American anomalies are too large, consistent with the model's excessive El Niño variability within the equatorial Pacific. The connection between the two regions is realistic if the model's tropical amplitude is accounted for. In the GISS model, anomalous rainfall during ENSO is small, consistent with the weak tropical wind stress anomaly cited above. The influence of El Niño over North America is nearly negligible in this model. The weak rainfall anomaly presumably is a result of unrealistic coupling between atmospheric and ocean physics. When SST instead is prescribed in this model, rainfall calculated by the GISS ModelE AGCM over the American Southwest is significantly correlated with El Niño as observed.

Realistic simulation of El Niño and its global influence remains a challenge for coupled models because of myriad contributing processes and their changing importance in the observational record. Key aspects of coupling between ocean and atmosphere—the relation between SST and wind stress anomalies, for example—are the result of complicated interactions among resolved model circulations, along with parameterizations of ocean and atmospheric boundary layers and moist convection. Simple models identify parameters controlling the magnitude and frequency of El Niño, such as the wind anomaly resulting from a change in SST (e.g., Zebiak and Cane 1987; Fedorov and Philander 2000), offering guidance to improve the realism of fully coupled GCMs. However, in a GCM, the coupling strength is emergent rather than prescribed, and it is often unclear a priori how to

change the coupling. Nonetheless, improved simulations of the ENSO cycle compared to previous generations (AchutaRao and Sperber 2006) suggest that additional realism can be expected in the future.

5.2.2.6 ANNULAR MODES

The primary mode of Arctic interannual variability is the Arctic Oscillation (Thompson and Wallace 1998), which also is referred to as the northern annular mode (NAM) and is related to the North Atlantic Oscillation (Hurrell 1995). The primary mode of Antarctic interannual variability is the southern annular mode (SAM) (Thompson and Wallace 2000), also known as Antarctic Oscillation. The variability modes are particularly important for attributing and projecting climate change; observed circulation changes in the past few decades (especially in the Southern Hemisphere) and model-projected changes in future circulation strongly resemble these structures.

Coupled climate models have shown skill in simulating NAM (Fyfe, Boer, and Flato 1999; Shindell et al. 1999; Miller, Schmidt, and Shindell 2006). In some cases, too much variability in the simulation of sea-level pressure is associated with NAM (Miller, Schmidt, and Shindell 2006). Global climate models also realistically simulate SAM (Fyfe, Boer, and Flato 1999; Cai, Whetton, and Karoly 2003; Miller, Schmidt, and Shindell 2006), although some details of SAM (e.g., amplitude and zonal structure) show disagreement among global climate model simulations and reanalysis data (Raphael and Holland 2006; Miller, Schmidt, and Shindell 2006).

In response to increasing concentrations of greenhouse gases and tropospheric sulfate aerosols in the 20th Century, the multimodel average exhibits a positive trend in the annular mode index in both hemispheres, with decreasing sea-level pressure over the poles and a compensating increase in midlatitudes most apparent in the Southern Hemisphere (Miller, Schmidt, and Shindell 2006). A variety of modeling studies also have shown that trends in stratospheric climate can affect the troposphere's annular modes (Shindell et al. 1999). Indeed, an important result from atmospheric modeling in recent years is the realization that the stratospheric ozone hole has contributed significantly to observed trends in surface winds and sea-level pressure distribution in the Southern Hemisphere (Thompson and Solomon 2002; Gillett and Thompson 2003). The models, however, may not be trustworthy in their simulation of the relative magnitude of greenhouse gas and stratospheric ozone effects on the annular mode. They also may underestimate the coupling of stratospheric changes due to volcanic aerosols with annular surface variations (Miller, Schmidt, and Shindell 2006; Arblaster and Meehl 2006).

5.2.2.7 OTHER MODES OF MULTIDECADAL VARIABILITY

In the Arctic during the last century, two long-period warm events occurred, one between 1920 and 1950 and another beginning in the late 1970s. Wang et al. (2007) evaluated a set of CMIP3 models for their ability to reproduce the amplitudes of air temperature variability of this character. As examples, CCSM3 and GFDL-CM2 models contain variance similar to that observed in the Arctic region.

Multidecadal variability in the North Atlantic is characterized by the Atlantic Multidecadal Oscillation (AMO) index, which represents a spatial average of SST (Enfield, Mestas-Nuñez, and Trimble 2001). Kravtsov and Spannagle (2007) analyzed SST from a set of current generation climate models. Their analysis attempts to separate variability associated with internal ocean fluctuations from that associated with changes by anthropogenic contributions. By isolating the multidecadal period of several regions in the ensemble SST series through statistical methods, they found that models obtain the observed magnitude of the AMO (Kravtsov and Spannagle 2007).

In the midlatitude Pacific region, decadal variability generally is underrepresented in the ocean (e.g., volume transports as described by Zhang and McPhaden 2006), with some models approaching amplitudes seen in observations. Examination of complicated feedbacks between atmosphere and ocean at decadal and longer scales shows that, while climate models generally reproduce the SST pattern related to the Pacific Decadal Oscillation (PDO), observed correlations between PDO and tropical

SST are not seen in the models (e.g., Alexander et al. 2006).

One of the most difficult areas to simulate is the Indian Ocean because of the competing effects of warm water inflow through the Indonesian archipelago, ENSO, and monsoons. The processes interact to varying degrees, challenging a model's ability to simulate all system aspects with observed relative emphasis. An index used to understand variability is the Indian Ocean Dipole pattern that combines information about SST and wind stress fields (Saji et al. 1999). While most models evaluated by Saji, Xie, and Yamagata (2005) were able to simulate the Indian Ocean's response to local atmospheric forcing in short time periods (semiannual), longer-period events such as the ocean's response to ENSO changes in the Pacific were not simulated well.

5.2.3 Polar Climates

Changes in polar snow and ice cover affect the Earth's albedo and thus the amount of insolation heating the planet (e.g., Holland and Bitz 2003; Hall 2004; Dethloff et al. 2006). Melting glaciers and ice sheets in Greenland and western Antarctica could produce substantial sea-level rise (Arendt et al. 2002; Braithwaite and Raper 2002; Alley et al. 2005). Polar regions thus require accurate simulation for projecting future climate change and its impacts.

Polar regions present unique environments and, consequently, challenges for climate modeling. Key processes include sea ice, seasonally frozen ground, and permafrost (Lawrence and Slater 2005; Yamaguchi, Noda, and Kitoh 2005). Processes also include seasonal snow cover (Slater et al. 2001), which can have significant subgrid heterogeneity (Liston 2004), and clear-sky precipitation, especially in the Antarctic (King and Turner 1997; Guo, Bromwich, and Cassano 2003). Polar regions test the ability of models to handle extreme geophysical behavior such as longwave radiation in clear, cold environments (Hines et al. 1999; Chiacchio, Francis, and Stackhouse 2002; Pavolonis, Key, and Cassano 2004) and cloud microphysics in the relatively clean polar atmosphere (Curry et al. 1996; Pinto, Curry, and Intrieri 2001; Morrison and Pinto 2005). In addition, polar atmospheric boundary layers can be very stable (Duynkerke and de Roode 2001; Tjernström, Zagar, and Svensson 2004; Mirocha, Kosovic, and Curry 2005), and their simulation remains an important area for model improvement.

For polar regions, much of simulated-variability assessment has focused on primary modes of polar interannual variability, along with the northern and southern annular modes. Less attention has been given to the ability of global climate-system models to simulate shorter-duration climate and weather variability in polar regions. Uotila et al. (2007) and Cassano et al. (2007) evaluated the ability of an ensemble of 15 global climate-system models to simulate daily variability in sea-level pressure in the Antarctic and Arctic. In both polar regions, they found that the ensemble was not able to reproduce many features of daily synoptic climatology, with only a small subset of models accurately simulating the frequency of primary synoptic weather patterns identified in global reanalysis datasets. U.S. models discussed in detail in Chapter 2 of this report spanned the same range of accuracy as non-U.S. models, with GFDL and CCSM models part of a small, accurate subset. More encouraging results were obtained by Vavrus et al. (2006), who assessed the ability of seven global climate models to simulate extreme cold-air outbreaks in the Northern Hemisphere.

Attention also has been given to the ability of regional climate models to simulate polar climate. In particular, the Arctic Regional Climate Model Intercomparison Project (ARCMIP) engaged a suite of Arctic regional atmospheric models to simulate a common domain and period over the western Arctic (Curry and Lynch 2002). Rinke et al. (2006) evaluated spatial and temporal patterns simulated by eight ARCMIP models and found that the model ensemble agreed well with global reanalyses, despite some large errors for individual models. Tjernstrom et al. (2005) evaluated near-surface properties simulated by six ARCMP models. In general, surface pressure, air temperature, humidity, and wind speed all were well simulated, as were radiative fluxes and turbulent momentum flux. The research group also found that turbulent heat flux was poorly simulated and that, over an entire annual cycle, the accumu-

lated turbulent heat flux simulated by models was many times larger than the observed turbulent heat flux (Fig. 5.6).

In global models, polar climate may be affected by errors in simulating other planetary regions, but much of the difference from observations and the uncertainty about projected climate change stem from current limitations in polar simulation. These limitations include missing or incompletely represented processes and poor resolution of spatial distributions.

As with other regions, model resolution affects simulation of important processes. In polar regions, surface distributions of snow depth vary markedly, especially when snow drifting occurs. Improved snow models are needed to represent such spatial heterogeneity (e.g., Liston 2004), which will continue to involve scales smaller than resolved for the foreseeable future. Frozen ground, whether seasonally frozen or occurring as permafrost, presents additional challenges. Models for permafrost and seasonal soil freezing and thawing are being implemented in land surface models (see Chapter 2). Modeling soil freeze and thaw continues to be a challenging problem as characteristics of energy and water flowing through soil affect temperature changes. Such fluxes are poorly understood (Yamaguchi, Noda, and Kitoh 2005).

Frozen soil affects surface and subsurface hydrology, which influences the surface water's spatial distribution with attendant effects on other parts of the polar climate system such as carbon cycling (e.g., Gorham 1991; Aurela, Laurila, Tuovinen 2004), surface temperature (Krinner 2003), and atmospheric circulation (Gutowski et al. 2007). The flow of fresh water into polar oceans potentially alters their circulation, too. Surface hydrology modeling typically includes, at best, limited representation of subsurface water reservoirs (aquifers) and horizontal flow of water both at and below the surface. These features limit the ability of climate models to represent changes in polar hydrology, especially in the Arctic.

Vegetation has been changing in the Arctic (Callaghan et al. 2004), and projected warming, which may be largest in regions where snow and ice cover retreat, may produce further changes in vegetation (e.g., Lawrence and Slater 2005). Current models use static distributions of vegetation, but dynamic vegetation models will be needed to account for changes in land-atmosphere interactions influenced by vegetation.

A key concern in climate simulations is how projected anthropogenic warming may alter land ice sheets, whose melting could raise sea levels substantially. At present, climate models do not include ice-sheet dynamics (see Chapter 2), and thus cannot account directly for ways in which ice sheets might change, possibly changing heat absorption from the sun and atmospheric circulation in the vicinity of ice sheets.

Distributions of snow, ice sheets, surface water, frozen ground, and vegetation have important spatial variation on scales smaller than the resolutions of typical contemporary climate models. This need for finer resolution may be satisfied by regional models simulating just a polar region. Because both northern and southern polar regions are within circumpolar atmospheric circulations (cf. Giorgi and Bi 2000 and Gutowski et al. 2007b), their coupling with other regions is more limited than in the case of midlatitude regions, which could allow polar-specific models that focus on Antarctic and Arctic processes, in part, to improve modeling of surface-atmosphere exchange processes (Fig. 5.6). Although each process above has been simulated in finer-scale, stand-alone models, their interactions as part of a climate system also need to be simulated and understood.

5.2.3.1 SEA ICE

Sea ice plays a critical role in the exchange of heat, mass, and momentum between ocean and atmosphere, and any errors in the sea-ice system will contribute to errors in other components. Two recent papers (Holland and Raphael 2006; Parkinson, Vinnikov, and Cavalieri 2006a, b) quantify how current models simulate the climate system's sea-ice process. Very limited observations make any evaluation of sea ice difficult. The primary observation available is sea-ice areal concentration. In some comparisons, sea-ice extent (the area where local ice concentration is greater than 15%) is used. For the past few decades, satellites have made it possible to produce a more complete dataset of

observations. Observations of ice extent were fewer before that. Other quantities that might be evaluated include ice thickness, but, due to limited observations, comparisons with models are difficult and will not be discussed further here.

The seasonal pattern in ice growth and decay in polar regions for all the models is reasonable (Holland and Raphael 2006; see Fig. 5.7). However, a large amount of variability between models occurs in their representation of sea-ice extent in both Northern and Southern hemispheres. Generally, models do better in simulating the Arctic than the Antarctic region, as shown with Fig. 5.8. An example of the complex nature of reproducing the ice field is given in Parkinson, Vinnikov, and Cavalieri (2006a,b), which found that all models showed an ice-free region in winter to the west of Norway, as seen in observational data, but all also produced too much ice north of Norway. The authors suggest that this is because the North Atlantic Current is not being simulated correctly. In a qualitative comparison, Hudson Bay is ice covered in winter in all models correctly reproducing the observations. The set of models having the most fidelity in the Arctic is not the same as the set having the most fidelity in the Antarctic. This difference may be due to distinctive ice regimes in the north and south or to differences in simulations of oceanic or meteorological circulations in those regions.

Holland and Raphael (2006) examined carefully the variability in Southern Ocean sea-ice extent. As an indicator of ice response to large-scale atmospheric events, they compared data from a set of IPCC AR4 climate models to the atmospheric index SAM for the April–June (AMJ) period (see Table 5.2). The models show that ice variability does respond modestly to large-scale atmosphere forcing but less than the limited observations show. Table 5.2 uses the U.S. models to examine whether models exhibit the observed out-of-phase buildup of ice between the Atlantic and Pacific sectors (referred to as the Antarctic Dipole).

Figure 5.6. Cumulative Fluxes of Surface Sensible Heat (top panel) and Latent Heat (bottom) at the SHEBA Site.

Data are from six models simulating a western Arctic domain for Sept. 1997 through Sept. 1998 for ARCMIP. SHEBA observations are gray shaded regions; model results are shown by the individual curves identified in the key at the lower left of the upper panel. [Figure adapted from Fig. 10(c and d) in M. Tjernstrom et al. 2005: Modelling the Arctic boundary layer: An evaluation of six ARCMIP regional-scale models with data from the SHEBA project. *Boundary-Layer Meteorology*, **117**, 337–381. Reproduced with kind permission of Springer Science and Business Media.]

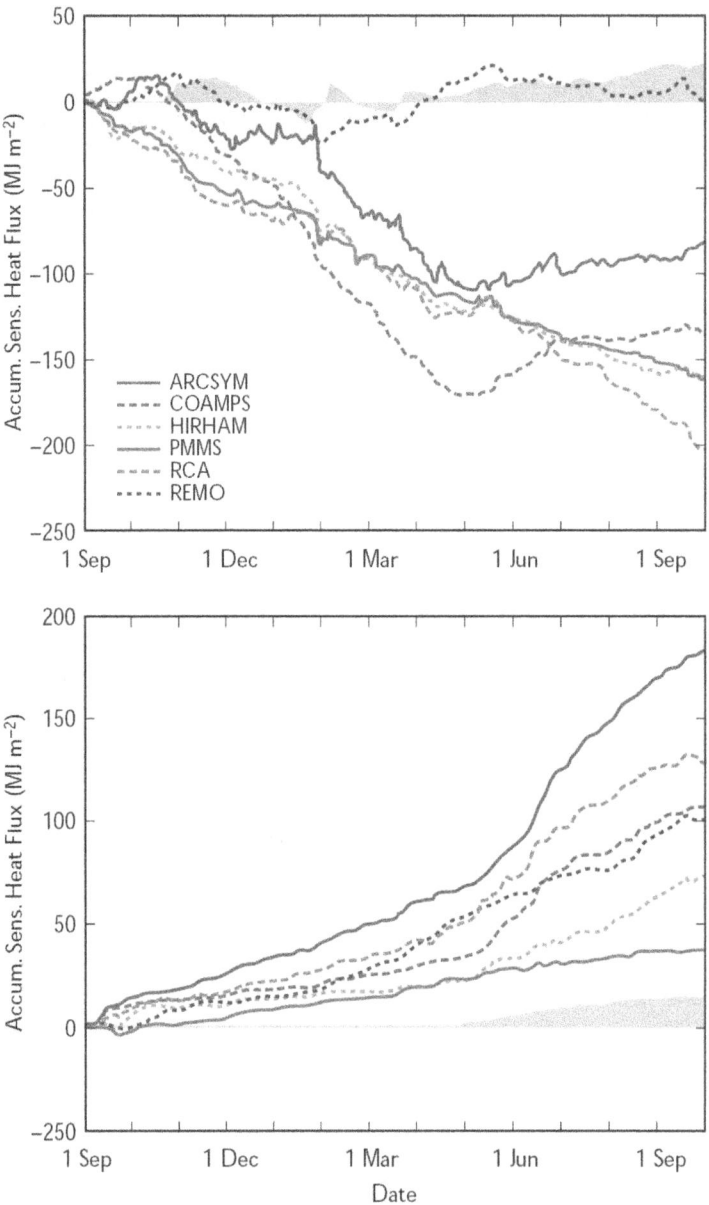

Table 5.2. Correlations of the Leading Mode of Sea-Ice Variability and Southern Annular Mode (SAM) for Observations and Model Simulations

	AMJ SAM and High-Pass Filtered Fields	AMJ SAM and Detrended Fields
Observations	**0.47**	**0.47**
CCSM3	**0.40**	**0.44**
GFDL-CM2.1	**0.39**	0.19
GISS-ER	0.30	0.20

The leading mode of sea-ice variability represents a shift of ice from the Atlantic to the Pacific sector. Bold values are significant at the 95% level, accounting for autocorrelation of the time series.

[Table modified from Table 1, p. 19, in M.M. Holland and M.N. Raphael 2006: Twentieth Century simulations of the Southern Hemisphere climate in coupled models. Part II: Sea ice conditions and variability. *Climate Dynamics*, **26**, 229–245. Reproduced with kind permission of Springer Science and Business Media.]

SH Total Extent

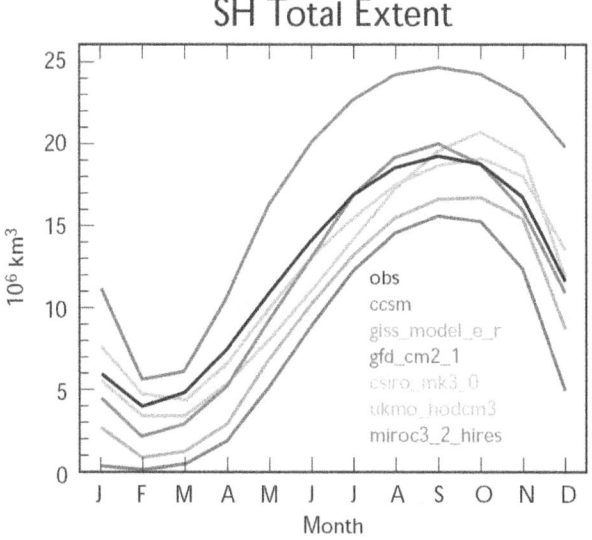

obs
ccsm
giss_model_e_r
gfd_cm2_1
csiro_mk3_0
ukmo_hadcm3
miroc3_2_hires

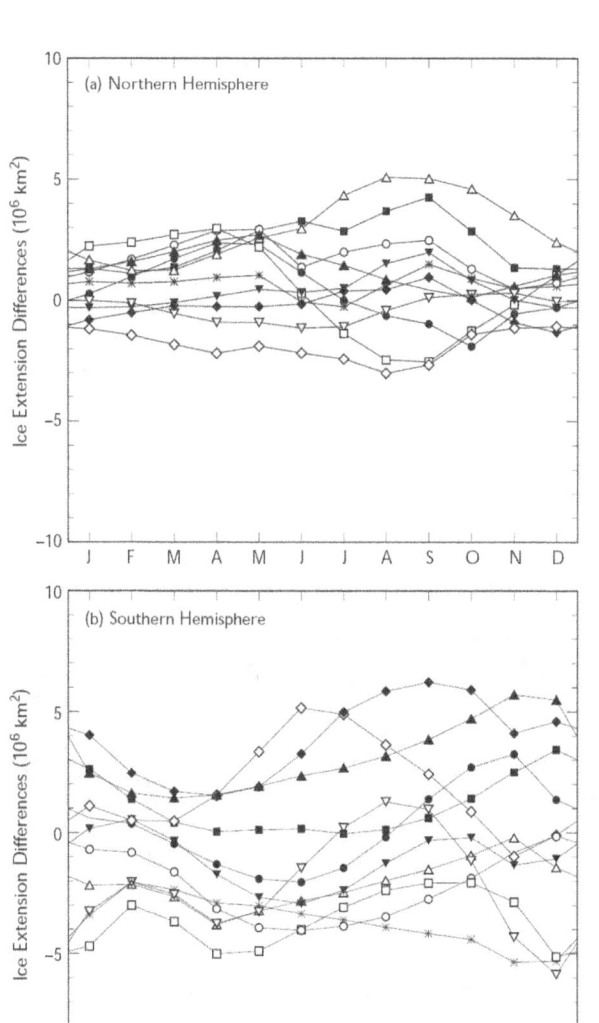

Figure 5.7. Annual Cycle of Southern Hemisphere Ice Extent.

It is defined as the area of ice with concentrations greater than 15%. Observations are identified by the black curve labeled "Obs," while the results from individual models are identified by the six colored curves. [From Fig. 1 in M.M. Holland and M.N. Raphael 2006: Twentieth Century simulations of the Southern Hemisphere climate in coupled models. Part II: Sea ice conditions and variability. *Climate Dynamics*, **26**, 229–245. Reproduced with kind permission of Springer Science and Business Media.]

Figure 5.8. Difference Between Modeled 1979 to 2004 Monthly Average Sea-Ice Extents and Satellite-Based Observations (modeled minus observed).

Data are shown for each of 11 major GCMs for both (a) Northern Hemisphere and (b) Southern Hemisphere. [From Fig. 4 in C.L. Parkinson, K.Y. Vinnikov, and D.J. Cavalieri 2006: Correction to evaluation of the simulation of the annual cycle of Arctic and Antarctic. *J. Geophysical Research*, **111**, C07012. Reproduced by permission of the American Geophysical Union (AGU).]

HadCM3
HadGEM1
ECHAM5
CGCM3
CSIRO Mk3
MIROC3
BCCR BCM2
GISS ER
IPSL CM4
INM CM3
GFDL CM2.1

5.2.4 Ocean Structure and Circulation

Unlike the atmosphere, the amount of observational data available to evaluate ocean simulations is very limited for long time periods. Nevertheless, sufficient data exist to identify a set of ocean characteristics or metrics to evaluate ocean models for their climate simulation properties. The most important is sea-surface temperature, but other quantities that serve as good indicators of ocean realism in climate models are ocean heat uptake, meridional overturning and ventilation, sea-level variability, and global sea-level rise.

5.2.4.1 SEA-SURFACE TEMPERATURE

Sea-surface temperature (SST) plays a critical role in determining climate and the predictability of climate changes. Because of interactions in atmospheric and ocean circulations at the sur-

face, errors in SSTs typically originate with deficiencies in both atmospheric and ocean model components. In general, more recent model versions show improvement over previous models when simulated SST fields are compared to observations. Figure 5.9 (Delworth et al. 2006) shows comparisons of simulated and observed mean SST fields of both the older GFDL CM2.0 and newer CM2.1 averaged over a 100-year period. The new model reduced a cold bias in the Northern Hemisphere from earlier simulations, resulting in both a more-realistic representation of atmospheric wind stress at the ocean surface and a modified treatment of sub-grid-scale oceanic mixing. The CCSM3.0 model's improved SST simulation over CCSM2.0 results mainly from changes in representing processes associated with the mixed layer of upper ocean waters (Danabasoglu et al. 2006).

Figure 5.9. Maps of Simulation Errors in Annual Mean SST.

Units are Kelvin (K). Errors are computed as model minus observations from Reynolds SST data (provided by NOAA-CIRES Climate Diagnostics Center, Boulder, Colorado, from their Web site, www.cdc.noaa.gov). (a) CM2.0 (using model years 101 to 200). (b) CM2.1 (using model years 101 to 200). Contour interval is 1 K, except for no shading of values between 1 K and +1 K. [Images from T.L. Delworth et al. 2006: GFDL's CM2 global coupled climate models. Part 1: Formulation and simulation characteristics. *J. Climate*, **19**, 643–684. Reproduced by permission of the American Meteorological Society.]

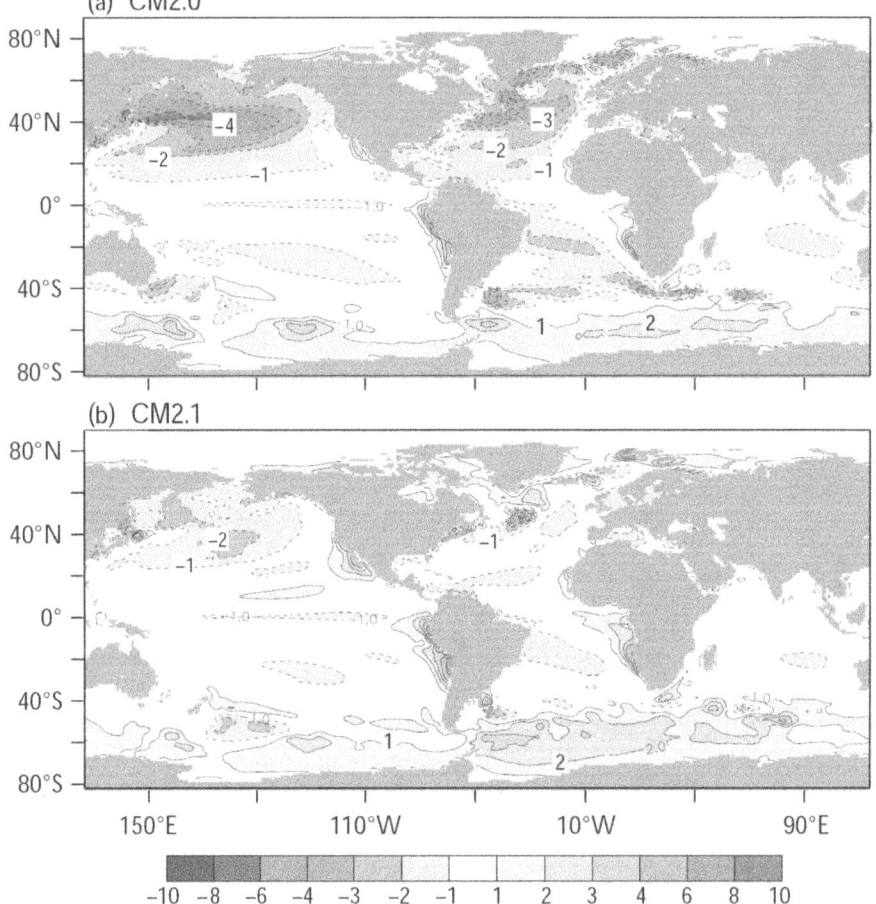

In addition to SST mean values, 20th Century trends of SST changes also are significant for model evaluation, since ocean SST contributes the dominant signal to the observed global surface temperature trend. An intermodel comparison of 50-year tropical SST trends is shown in Fig. 5.10. Trends range from a low of 0.1°C/50 yrs to a high of about 0.6°C/50 yrs, with the observational trend estimate given as about 0.43°C/50 yrs. The figure also shows some randomness within a group of simulations run by the same model. For example, the two different GFDL model versions discussed above were each run for multiple realizations of the 20th Century. CM 2.0 simulations are noted by GFDL201, GFDL202, and GFDL203, and CM 2.1 simulations are noted by GFDL211, GFDL212, and GFDL213.

5.2.4.2 MERIDIONAL OVERTURNING CIRCULATION AND VENTILATION

The planetary-scale circulation transporting heat and freshwater throughout global oceans is referred to as global thermohaline circulation. The Atlantic portion is called the Atlantic meridional overturning circulation (AMOC). Tropical and warm waters flow northward via the Gulf Stream and North Atlantic Current. Southward flow occurs when water is subducted in regions around Labrador and Greenland; surface waters freshen, become denser, and flow down the slope to deeper depths. Similar processes occur at locations in the Southern Ocean. "Ventilation" is the name given to the process by which these dense surface waters are carried into the ocean interior. An important climate parameter is the rate at which this process occurs. The pattern of circulation may weaken,

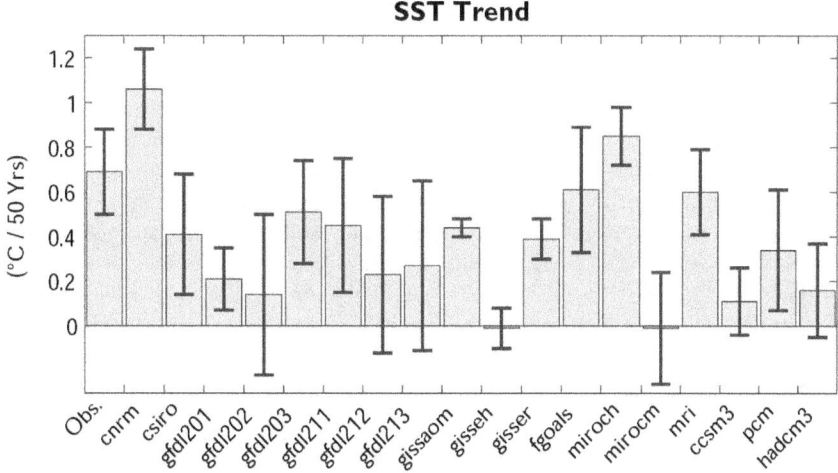

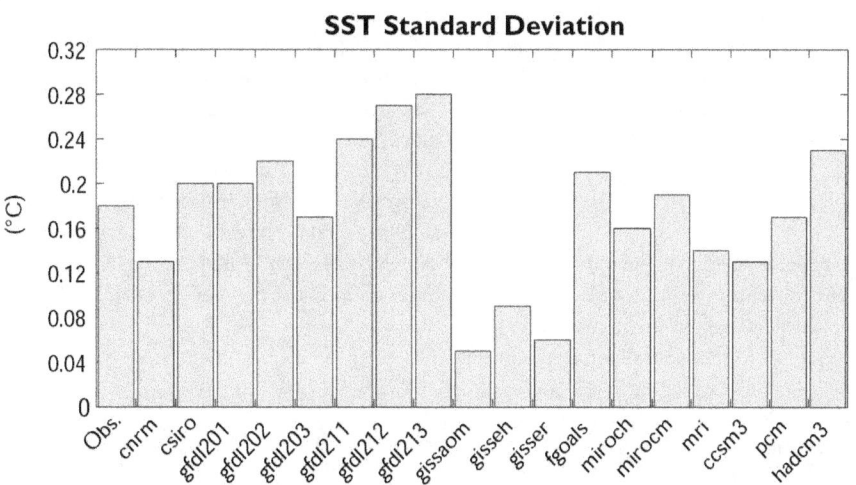

Figure 5.10. Trends and Standard Deviations of Tropical SST Between 1950 and 1999.

Observations are shown by the leftmost bar in each figure. All others are model results. Error bars show 95% significance levels for trends. [Images from Fig. 9 in D. Zhang and M.J. McPhaden 2006: Decadal variability of the shallow Pacific meridional overturning circulation: Relation to tropical sea-surface temperatures in observations and climate change models. *Ocean Modelling*, **15**, 250–273. Used with permission from Elsevier.]

affecting the climate in the region surrounding the North Atlantic. Schmittner, Latif, and Schneider (2005) examined a small ensemble set of simulations to quantify uncertainty in model representation of 20th Century AMOC transports. To make their estimate, they evaluated global temperature, global salinity, pycnocline depth, surface temperature, surface salinity in the Atlantic (SST, SSS), and the overturning calculations at three Atlantic locations. Their results suggest that temperature is simulated most successfully on a large scale and that the overturning transports at 24°N are close (~18 Sv) to observed measurements (~15.8 Sv). However, the maximum mean overturning transports in these models are too high, between 21.2 and 31.7 Sv, when compared to the observed value (17.7 Sv). Several other CMIP3 models underestimated maximum transport. The authors do not attempt to explain why models are different from each other and from observations.

Another aspect of planetary-scale ocean circulation of interest is transport of mass by the Antarctic Circumpolar Current through the Drake Passage. The passage, between the tip of South America and the Antarctic Peninsula, provides a constrained passage to measure the flow between two large ocean basins. Observed mean transport is around 135 Sv (Cunningham et al. 2003). Russell, Stouffer, and Dixon (2006, 2007) estimate passage flow for a subset of climate models. Simulated mean values show a wide range. For example, GFDL and GISS-EH models do fairly well in reproducing the observed average transport with values between 113 and 175 Sv. Once again, the interaction between the atmospheric and ocean component models appears to be important in reproducing the observed transport. The strength and location of the zonal wind stress provided by the atmosphere correlate with how well the transport reflects observed values.

5.2.4.3 NORTHWARD HEAT TRANSPORT

A common metric used to quantify the realism in ocean models is the northward transport of heat. This integrated quantity (from top to bottom and across latitude bands) gives an estimate of how heat moves within the ocean and is important in balancing the overall heat exchange between the tropics and the extratropical regions

of the Earth. The calculations for the ocean's northward heat transport in the current generation of climate models show that the models reasonably represent the observations (Delworth et al. 2006; Collins et al. 2006a; Schmidt et al. 2006). The current models have significantly improved over the last generation in the Northern Hemisphere. Comparisons of simulated values to observed values for the North Atlantic are within the uncertainty of the observations. In the Southern Hemisphere, the comparisons in all the models are not as good, with the Indian Ocean transport estimates contributing to a significant part of the mismatch. In coupled ocean-atmosphere simulations, erroneous ocean heat transport is compensated by changes in atmospheric heat transport that give a more realistic total heat transport (Covey and Thompson 1989).

Heat Content. The global mean mass-weighted ocean temperature is called the ocean's heat content. Its time evolution is centrally important in determining how realistically the models reproduce heat uptake. The seasonal cycle and longer-term trends of heat content provide useful model metrics, although the seasonal cycle does not affect the deep ocean. An evaluation of temporally evolving ocean-heat content in the CMIP3 suite of climate models shows the models' abilities to simulate the zonally integrated annual and semiannual cycle in heat content. In the middle latitudes (Gleckler, Sperber, and AchutaRao 2006), the models do a reasonable job, although a broad spread of values is apparent for tropical and polar regions. This analysis showed that the models replicate the annual cycle's dominant amplitude along with its phasing in the midlatitudes (Figs. 5.11 a–b and 5.12 a–f). At high latitudes, comparisons with observations are not as consistent. Although the annual cycle and global trend are reproduced, model analyses (e.g., Hansen et al. 2005a, b) show they do not simulate decadal changes in estimates made from observations (Levitus et al. 2001). Part of the difficulty of comparisons at high latitudes and long periods is the paucity of observational data (Gregory et al. 2004).

a)

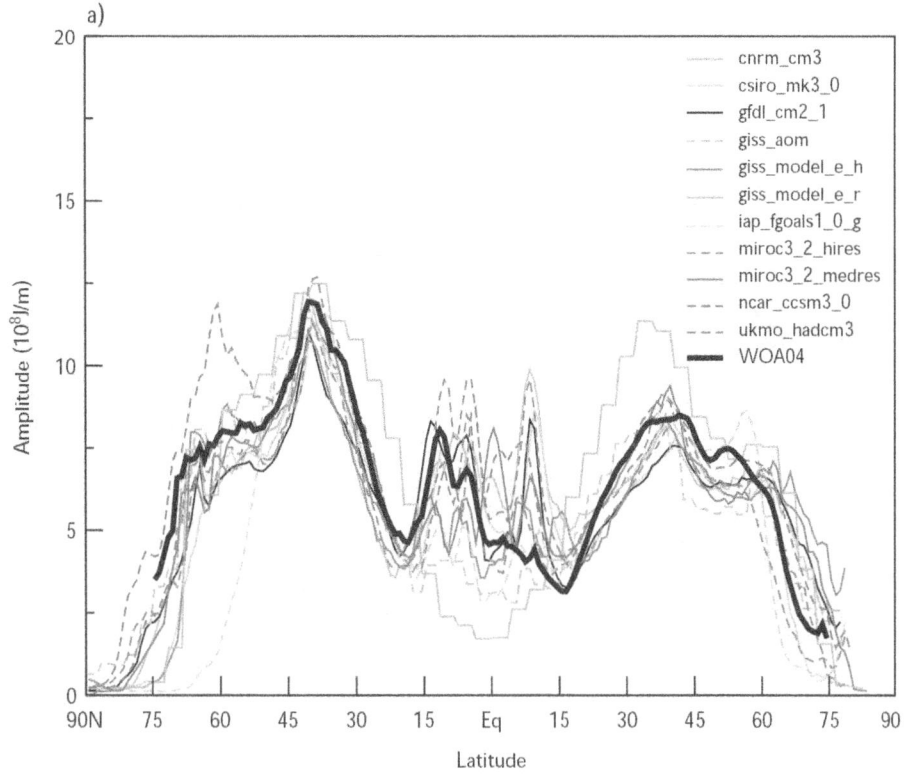

**Figure 5.11a–b.
Observed and
Simulated Zonally
Integrated Ocean
Heat Content
(0–250 m).**

Observations are
represented by the curve
labeled "WOA04." All
other curves are model
results. (a) annual cycle
amplitude (10^8 J/m²) and
(b) semiannual/annual
(A2/A1). [From Fig. 1 in P.J.
Gleckler, K.R. Sperber, and
K. AchutaRao 2006: Annual
cycle of global ocean heat
content: Observed and
simulated. *J. Geophysical
Research*, **111**, C06008.
Reproduced by
permission of the
American Geophysical
Union (AGU).]

b)

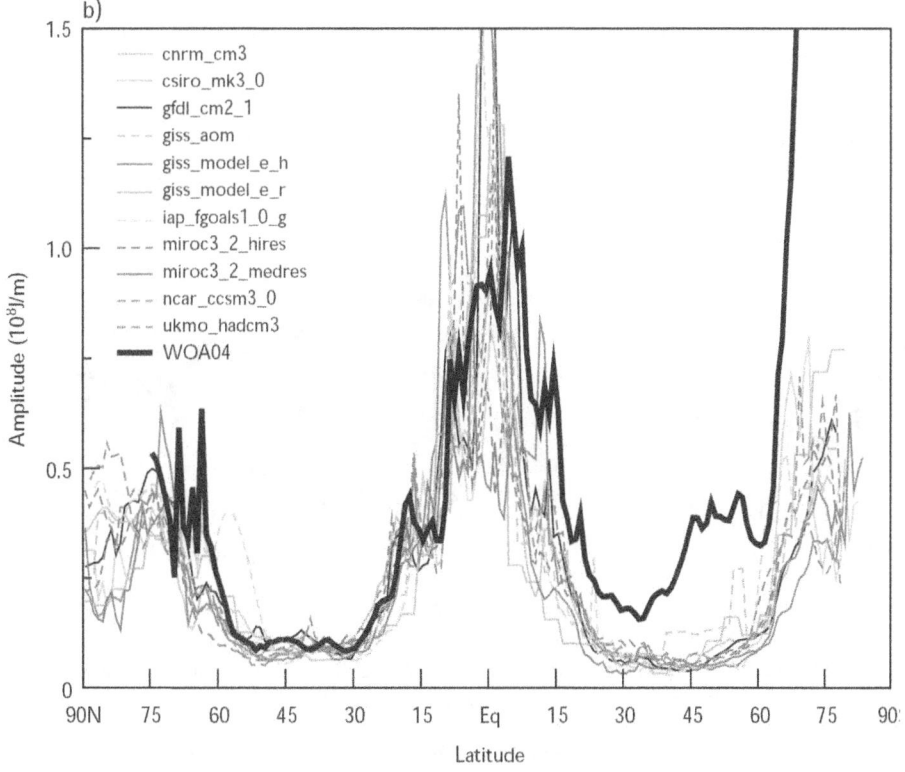

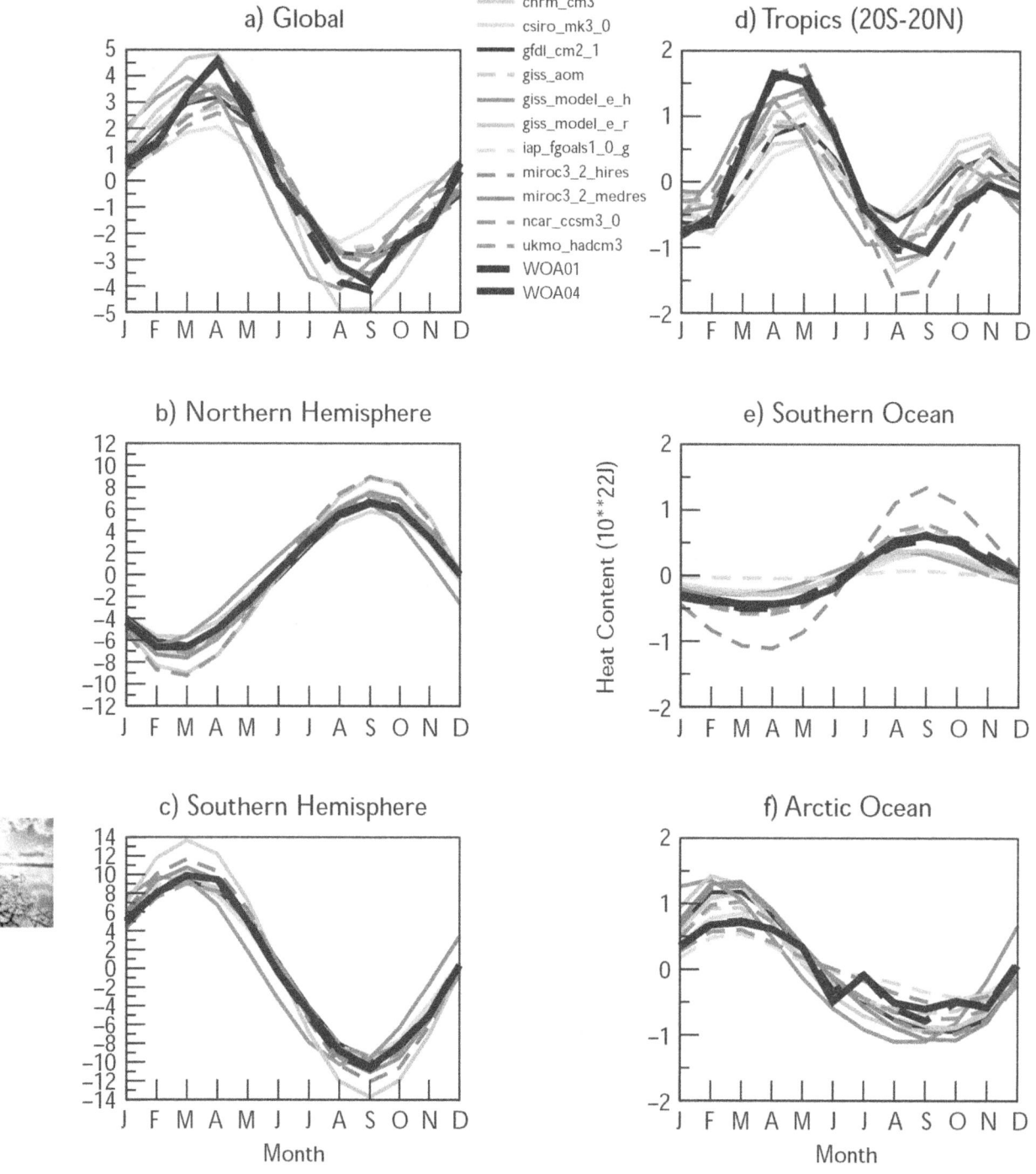

Figure 5.12a–f. Annual Cycle of Observed and Simulated Basin Average Global Ocean Heat Content (0–250 m).

Observations are represented by the curves labeled "WOA01" and "WOA04." Units are 10^{22} J. Arctic Ocean is defined as north of 60°N, and Southern Ocean is south of 60°S. [From Fig. 3 in P.J. Gleckler, K.R. Sperber, and K. AchutaRao 2006: Annual cycle of global ocean heat content: Observed and simulated. *J. Geophysical Research*, **111**, C06008. Reproduced by permission of the American Geophysical Union (AGU).]

5.2.5 Global Mean Sea-Level Rise

Two separate physical processes contribute to sea-level rising: (1) ocean thermal expansion from an increase in ocean heat uptake (steric component) and (2) addition of freshwater from precipitation, continental ice melt, and river runoff (eustatic component). Various ocean models handle freshwater fluxes in different ways. With the addition of a free surface in the current generation of ocean models, freshwater flux into oceans can be included directly (Griffies et al. 2001). The freshwater contribution is computed in quantities estimated by the climate model's atmosphere and ice-sheet components (e.g., Church, White, and Arblaster 2005; Gregory, Lowe, and Tett 2006). In general, state-of-the-art climate models underestimate the combined global mean sea-level rise as compared to tide gauge and satellite altimeter estimates, while the rise for each separate component is within the observed values' uncertainty. The reason for this is an open research question and may relate either to observational sampling or to incorrectly accounting for all eustatic contributions. The steric component to global mean sea-level rise is estimated at 0.40 ± 0.05 mm/yr from observations (Antonov, Levitus, and Boyer 2005). Models simulate a similar but somewhat smaller rise (Gregory, Lowe, and Tett 2006; Meehl et al. 2005). Significant differences also occur in the magnitudes of decadal variability between observed and simulated sea level. Progress is being made, however, over the previous generation of climate models. When atmospheric effects from volcanic eruptions are included, for example, current-generation ocean models capture the volcanoes' observed impact (a decrease in the global mean sea level). Figure 5.13 from Church, White, and Arblaster (2005) gives an example of a few models and their detrended estimate of the historic global mean sea level. It shows the influence of including additional atmospheric forcing agents in changing the ocean's steric height.

5.3 EXTREME EVENTS

Flood-producing precipitation, drought, heat waves, and cold waves have severe impacts on North America. Flooding resulted in average annual losses of $3.7 billion between 1983 and 2003 (www.flooddamagedata.org). Losses from the 1988 drought were estimated at $40 billion and the 2002 drought at $11 billion. Heat waves in 1995 resulted in 739 additional deaths in Chicago alone (Whitman et al. 1997). A large component of overall climate change impacts probably will arise from changes in the intensity and frequency of extreme events.

Modeling of extreme events poses special challenges since they are, by definition, rare. Although the intensity and frequency of extreme events are modulated by ocean and land surface state and by trends in the mean climate state, internal atmospheric variability plays a very large role, and the most extreme events arise from chance confluence of unlikely conditions. The very rarity of extreme events makes statistical evaluation of model performance less robust than for mean climate. For example, in evaluating a model's ability to simulate heat waves as intense as that in 1995, only a few episodes in the entire 20th Century approach or exceed that intensity (Kunkel et al. 1996). For such rare events, estimates of the real risk are highly uncertain, varying from once every 30 years to once every 100 years or more. Thus, a model that simulates these occurrences at a frequency of once every 30 years may be performing adequately, but its performance cannot be distinguished from that of the model that simulates a frequency of once every 100 years.

Although it might be expected that a change in mean climate conditions will apply equally to changes in extremes, this is not necessarily the case. Using as an example the 50-state record-low temperatures, the decade with the largest number of records is the 1930s, yet winters during that decade averaged third warmest since 1890; in fact, no significant correlation is shown between the number of records and U.S. wintertime temperature (Vavrus et al. 2006). Thus, the severest cold air outbreaks in the past do not necessarily coincide with cold winters. Another examination of model data showed that future changes in extreme temperatures differ from changes in mean temperature in many regions (Hegerl et al. 2004). This means that climate model output must be analyzed explicitly for extremes by examining daily (or even finer-resolution) data, a resource-intensive effort.

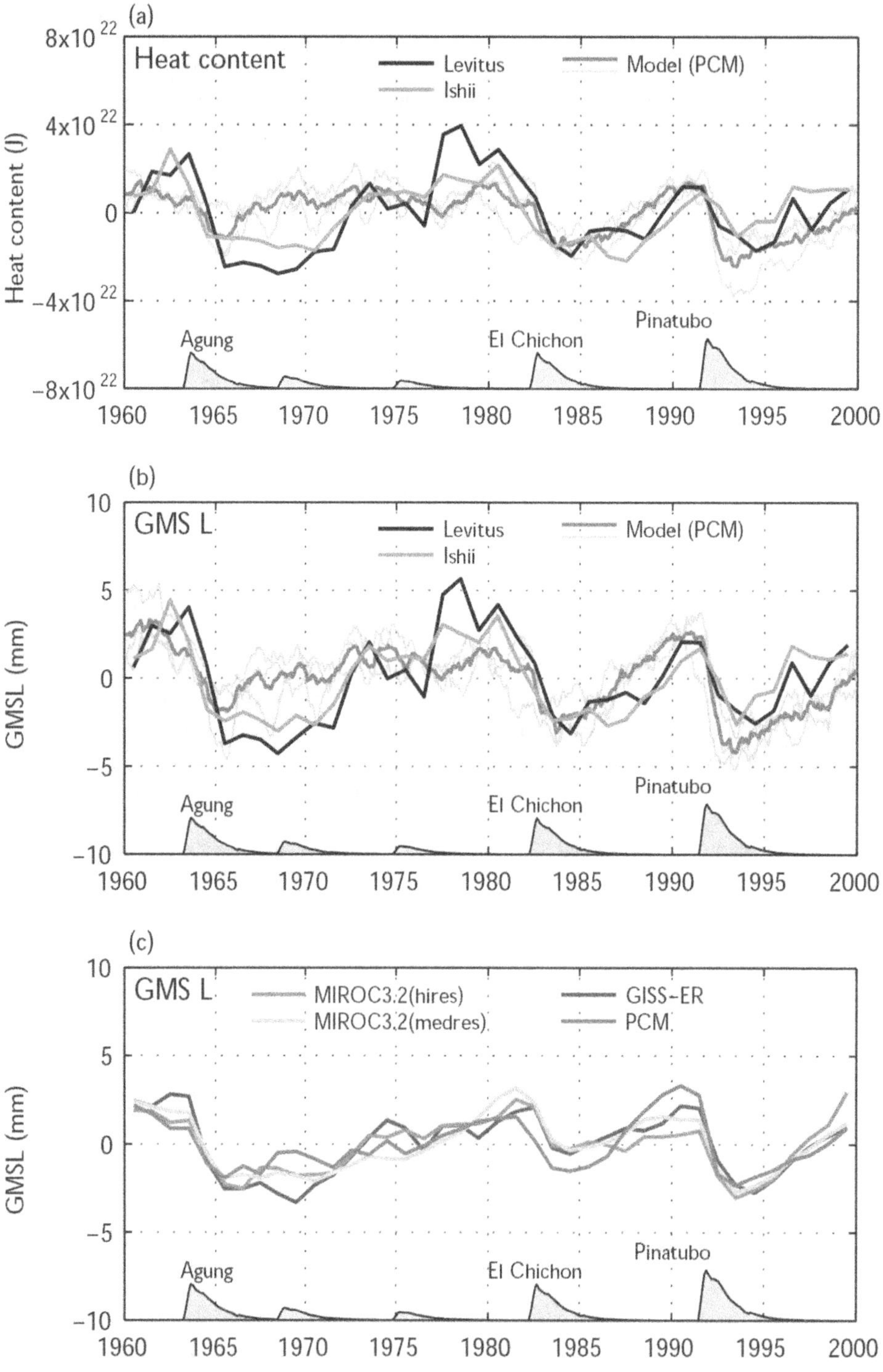

Figure 5.13. Observed and Modeled Global Ocean Heat Content (GOHC) and Global Mean Sea Level (GMSL) for 1960 to 2000.

The response to volcanic forcing, as indicated by differences between pairs of PCM simulations for GOHC (a) and GMSL (b) is shown for the ensemble mean (bold line) and the three ensemble members (light lines). Observational estimates of GOHC and GMSL are shown by the black and blue bold lines. For a and b, all results are for the upper 300 m only and have been detrended over the period 1960 to 2000. For c, the ensemble mean (full-depth) GMSL for GISS-ER, MIROC3.2(hires), MIROC3.2(medres), and PCM models (after subtracting a quadratic) are shown. [From Fig. 2 in J.A. Church, N.J. White, and M. Arblaster 2005: Significant decadal-scale impact volcanic eruptions on sea level and ocean heat content. *Nature*, **438**(7064), 74–77. Used with permission from Nature Publishing Group.]

Evaluation of model performance with respect to extremes is hampered by incomplete data on historical frequency and severity of extremes. Frich et al. (2002) analyzed ten indicators of climate extremes and presented global results. However, many areas were missing due to lack of suitable station data, particularly in the tropics. Using some of these indices for comparisons between models and observations has become common. Another challenge for model evaluation is the spatially averaged nature of model data, representing an entire grid cell, while station data represent point observations. For some comparisons, averaging station data over areas representing a grid cell is necessary.

Several approaches are used to evaluate model performance for simulation of extremes. One approach examines whether a model reproduces the magnitude of extremes. For example, a daily rainfall amount of 100 mm or more is expected to occur about once every year in Miami, every 6 years in New York City, every 13 years in Chicago, and every 200 years in Phoenix. A useful metric would be the extent to which a model is able to reproduce absolute magnitudes and spatial variations of such extremes. A second approach examines whether a model reproduces observed trends in extremes. Perhaps the most prominent observed global trend is an increase in the frequency of heavy precipitation, particularly during the last 20 to 30 years of the 20th Century. This trend is significant at the 95% confidence level for the period 1979 to 2003 and at the 99% confidence level for the period 1951 to 2003 (Trenberth et al. 2007). Another notable observed trend is an increase in the length of the frost-free season.

In some key respects, model simulation of temperature extremes probably is less challenging than simulating precipitation extremes, in large part due to the scales of these phenomena. The typical heat wave or cold wave covers a relatively large region, on the order of several hundred miles or more or a number of grid cells in a modern climate model. By contrast, heavy precipitation can be much more localized, often extending over regions of much less than 150 km, or less than the size of a grid cell. Thus, the modern climate model can simulate directly the major processes causing temperature extremes while heavy precipitation is sensitive to para-

meterization of subgrid-scale processes, particularly convection (Chapter 2; Emori and Brown 2005; Iorio et al. 2004).

5.3.1 Droughts and Excessive Rainfall Leading to Floods

Recent analysis indicates a globally averaged trend toward greater areal coverage of drought since 1972 (Dai et al. 2004). A simulation by the HadCM3 model reproduces this dry trend (Burke, Brown, and Christidis 2006) only if anthropogenic forcing is included. A control simulation indicates that the observed drying trend is outside the range of natural variability. The model, however, does not always correctly simulate the regional distributions of areas of increasing wetness and dryness. The relationship between droughts and variability was covered above in Section 5.2.2.3 Monsoons.

Several different measures of excessive rainfall have been used in analyses of model simulations. A common one is the annual maximum 5-day precipitation amount, one of the Frich et al. (2002) indices. This has been analyzed in several recent studies (Kiktev et al. 2003; Hegerl et al. 2004; Tebaldi et al. 2006). Other analyses have examined thresholds of daily precipitation, either absolute (e.g., 50 mm/day in Dai 2006) or percentile (e.g., 4th-largest precipitation event equivalent to 99th percentile of 365 daily values as in Emori et al. 2005). Recent studies of model simulations produced for CMIP3 provide information on the performance of the latest model generation.

Models generally tend to underestimate very heavy precipitation. This is shown in a comparison between satellite (TRMM) estimates of daily precipitation and model-simulated values within the 50°S–50°N latitude belt (Dai 2006). TRMM observations derive 7% of total precipitation from very heavy rainfall of 50 mm or more per day, in contrast to only 0 to 2% for the models. For the frequency of very heavy precipitation of 50 mm or more per day, TRMM data show a frequency of 0.35% (about once every 300 days), whereas it is 0.02 to 0.11% (once every 900 to 5000 days) for the models. A global analysis of model simulations showed that models produced too little precipitation in events exceeding 10 mm/day (Sun et al. 2006). Examining how many days it takes to accumulate two-thirds of annual precipitation, models generally show too many days compared to observations over North America, although a few models are close to reality. In contrast to the general finding of a tendency toward underestimation, a study (Hegerl et al. 2004) of two models indicates generally good agreement with observed annual maximum 5-day precipitation amounts over North America for HadCM3 and even somewhat of an overestimation for CGCM2.

This model tendency to produce rainfall events less intense than observed appears to be due in part to global models' low spatial resolution. Experiments with individual models show that increasing resolution improves the simulation of heavy events. For example, the fourth-largest precipitation event in a model simulation with a resolution of about 300 km averaged 40 mm over the conterminous United States, compared to an observed value of about 80 mm. When the resolution was increased to 75 km and 50 km, the fourth-largest event was still smaller than observed but by a much smaller amount (Iorio et al. 2004). A second important factor is the parameterization of convection. Thunderstorms are responsible for many intense events, but their scale is smaller than the size of model grids and thus must be indirectly represented in models (Chapter 2). One experiment showed that changes to this representation improve model performance and, when combined with high resolution of about 1.1° latitude, can produce quite-accurate simulations of the fourth-largest precipitation event on a globally averaged basis (Emori et al. 2005). Another experiment found that the use of a cloud-resolving model imbedded in a global model eliminated underestimation of heavy events (Iorio et al. 2004). A cloud-resolving model eliminates the need for convection parameterization but is very expensive to run. These sets of experiments indicate that the problem of heavy-event underestimation may be reduced significantly in future as increases in computer power allow simulations at higher spatial resolution and perhaps eventually the use of cloud-resolving models.

Improved model performance at higher spatial resolutions provides motivation for use of regional climate models when only a limited area,

such as North America, is of interest. These models have spatial resolution sufficient to resolve major mountain chains, and some thus display considerable skill in areas where topography plays a major role in spatial patterns. For example, they are able to reproduce rather well the spatial distribution of the magnitude or extent of precipitation in the 95th percentile (Leung and Qian 2003), frequency of days with more than 50 mm and 100 mm (Kim and Lee 2003), frequency of days over 25 mm (Bell, Sloan, and Snyder 2004), and annual maximum daily precipitation amount (Bell, Sloan, and Snyder 2004) over the western United States. Kunkel et al. (2002) found that an RCM's simulation of extreme-event magnitude over the United States varied spatially and depended on event duration. There was a tendency for overestimation in western United States and good agreement or underestimation in central and eastern United States.

Most studies of observed precipitation extremes suggest that they have increased in frequency and intensity during the latter half of the 20th Century. A study by Tebaldi et al. (2006) indicates that models generally simulate a trend toward a world characterized by intensified precipitation, with a greater frequency of heavy-precipitation and high-quantile events, although with substantial geographical variability. This is in agreement with observations. Wang and Lau (2006) find that CGCMs simulate an increasing trend in heavy rain over the tropical ocean.

5.3.2 Heat and Cold Waves

Analyses of simulations for IPCC AR4 by seven climate models indicate that they reproduce the primary features of cold air outbreaks (CAOs), with respect to location and magnitude (Vavrus et al. 2006). In the analyses, a CAO is an episode of at least 2 days duration during which the daily mean winter (December-January-February) surface temperature at a gridpoint is two standard deviations below the gridpoint's winter mean temperature. Maximum frequencies of about four CAO days per winter are simulated over western North America and Europe, while minimal occurrences of less than one day per winter exist over the Arctic, northern Africa, and parts of the North Pacific. GCMs generally are accurate in their simulation of primary fea-

tures, with high pattern correlation to observations and maximum number of days meeting CAO criteria around 4 per winter. One favored region for CAOs is in western North America, extending from southern Alaska into the upper Midwest. Here, models simulate a frequency of about 4 CAO days per year, in general agreement with the observed values of 3 to 4 days. Models underestimate frequency in the southeastern United States (mean simulated values range from 0.5 to 2 days vs 2 to 2.5 days in observations). This regional bias occurs in all models and reflects the inability of GCMs to penetrate Arctic air masses far enough southeastward over North America.

CMIP3 model simulations show a positive trend for growing season, heat waves, and warm nights and a negative trend for frost days and daily temperature range (maximum minus minimum) (Tebaldi et al. 2006). The simulations indicate that this is in general agreement with observations, except that there is no observed trend in heat waves. The modeled spatial patterns generally have larger positive trends in western North America than in eastern sections. For the United States, this is in qualitative agreement with observations showing that decreases in frost-free season and frost days are largest in the western United States (Kunkel et al. 2004; Easterling 2002).

Analysis of individual models provides a more detailed picture of model performance. In a simulation from PCM (Meehl, Tebaldi, and Nychka 2004), the largest trends for decreasing frost days occur in the western and southwestern United States (values greater than –2 days per decade). Trends near zero in the upper Midwest and northeastern United States show good agreement with observations. The biggest discrepancy between model and observations is over parts of the southeastern United States, where the model shows trends for decreasing frost days and observations show slight increases. This is thought to be a partial consequence of two large El Niño events in observations during this time period (1982–1983 and 1997–1998) when anomalously cool and wet conditions occurred over the southeastern United States and contributed to slight increases of frost days. The model's ensemble mean averages out effects from individual El

Niño events, and thus frost-day trends reflect a more general response to forcings that occurred during the latter part of the 20th Century. An analysis of short-duration heat waves simulated by PCM (Meehl and Tebaldi 2004) indicates good agreement with observed heat waves for North America. In that study, heat waves were defined by daily minimum temperature. The most intense events occurred in the southeastern United States for both model simulation and observations. The overall spatial pattern of heatwave intensity in the model matched closely with the observed pattern. In a four-member ensemble of simulations from HadCM3 (Christidis et al. 2005), the model showed a rather uniform pattern of increases in the warmest night for 1950 to 1999. Observations also show a global mean increase, but with considerable regional variations. In North America, observed trends in the warmest night vary from negative in the south-central sections to strongly positive in Alaska and western Canada, compared to a rather uniform pattern in the model. However, this discrepancy might be expected, since the observations probably reflect a strong imprint of internal climate variability that is reduced by ensemble averaging of the model simulations.

Analysis of the magnitude of temperature extremes for California in a regional climate model simulation (Bell, Sloan, and Snyder 2004) shows mixed results. The hottest maximum in the model is 4°C less than observations, while the coldest minimum is 2.3°C warmer. The number of days >32°C is 44 in the model compared to an observed value of 71. This could result from the lower diurnal temperature range in the model (15.4°C observed vs 9.7°C simulated). While these results are better than the driving GCM, RCM results are still somewhat deficient, perhaps reflecting the study region's very complex topography.

Models display some capability to simulate extreme temperature and precipitation events, but there are differences from observed characteristics. Models typically produce global increases in extreme precipitation and severe drought and decreases in extreme minimum temperatures and frost days, in general agreement with observations. Models have a general, though not universal, tendency to underestimate the magnitude of heavy precipitation events.

Regional-trend features are not always captured. Since the causes of observed regional-trend variations are not known in general and such trends could be due in part to the climate system's stochastic variability, assessing the significance of these discrepancies is difficult.

6 Future Model Development

CHAPTER

Climate models are evolving toward greater comprehensiveness, incorporating such aspects of the chemical and biological environment as active vegetation on land and oceanic biogeochemistry that affect and are affected by the physical climate. Climate models are simultaneously evolving toward finer spatial resolution.

Improvements in climate simulations as resolution increases can be both incremental and fundamental. Incremental improvements are expected in treatment of the atmosphere due to better simulation of atmospheric fronts, interactions among extratropical storms and sharp topographic features, and, especially, tropical storms. In the ocean, finer resolution incrementally improves the simulation of narrow boundary currents and the circulation in relatively small basins, such as the Labrador Sea, that play key roles in oceanic circulation.

More fundamental changes also happen in both the atmosphere and the ocean as resolution improves. In the ocean a key transition occurs at grid scales of tens of kilometers, at which point mesoscale eddies (see Chapter 2) begin to be explicitly resolved. In the atmosphere, a fundamental transition takes place when the grid scale drops to a few kilometers, where direct simulation of dominant deep convective circulations begins to be feasible and the model's dependence on uncertain subgrid-scale parameterization of deep moist convection diminishes.

In the following, we discuss these more fundamental oceanic and atmospheric transitions and then describe some examples of increased comprehensiveness in climate modeling (see also Chapter 2 for glacial modeling, another important future development).

The climate modeling enterprise is evolving along additional paths (apart from evolution of the models themselves) that are not discussed here. One path is the creation of large ensembles of model simulations by varying uncertain physical parameters so as to better estimate the associated uncertainties [quantifying uncertainty in model predictions (called QUMP); Murphy et al. 2004; climateprediction.net]. Others include the movement toward initializing climate models with estimates of observed climatic states, particularly the observed oceanic state, so as to optimize the realism of decadal forecasts, which marks an evolution toward the merging of seasonal-interannual and decadal forecasting (Troccoli and Palmer 2007).

6.1 HIGH-RESOLUTION MODELS

6.1.1 Mesoscale Eddy-Resolving Ocean Models

The distinction between laminar and turbulent flow in the ocean is fundamental. Simulations of the more realistic turbulent regime promise to substantially raise the level of realism in oceanic climate simulations. For example, Fig. 6.1 shows two simulations of the Southern Ocean by an ocean model developed at the Geophysical Fluid Dynamics Laboratory (GFDL) (Hallberg and Gnanadesikan 2006). The field shown is an instantaneous snapshot of the surface current speed. Resolution of the model on the left is about 1° latitude. The result is a relatively laminar (nonturbulent) flow with a gently meandering circumpolar current. The figure on the right is obtained by reducing the grid size to 1/6 of a degree. A much more turbulent flow is simulated by the model with abundant vortex generation. This model is beginning to resolve the spectrum of mesoscale eddies that populate the Southern Ocean and many other oceanic regions. As discussed in Chapter 2, the effects on ocean circulation of mesoscale eddy-induced mixing are parameterized in current ocean models, which can be thought of as essentially laminar.

While progress has been made in recent years, explicit simulation of these eddies undoubtedly is more reliable than mixing parameterizations. In the Southern Ocean, eddies are thought to control the circumpolar current's response to wind changes (Hallberg and Gnanadesikam 2006) and the way carbon dioxide is taken up by the Southern Ocean.

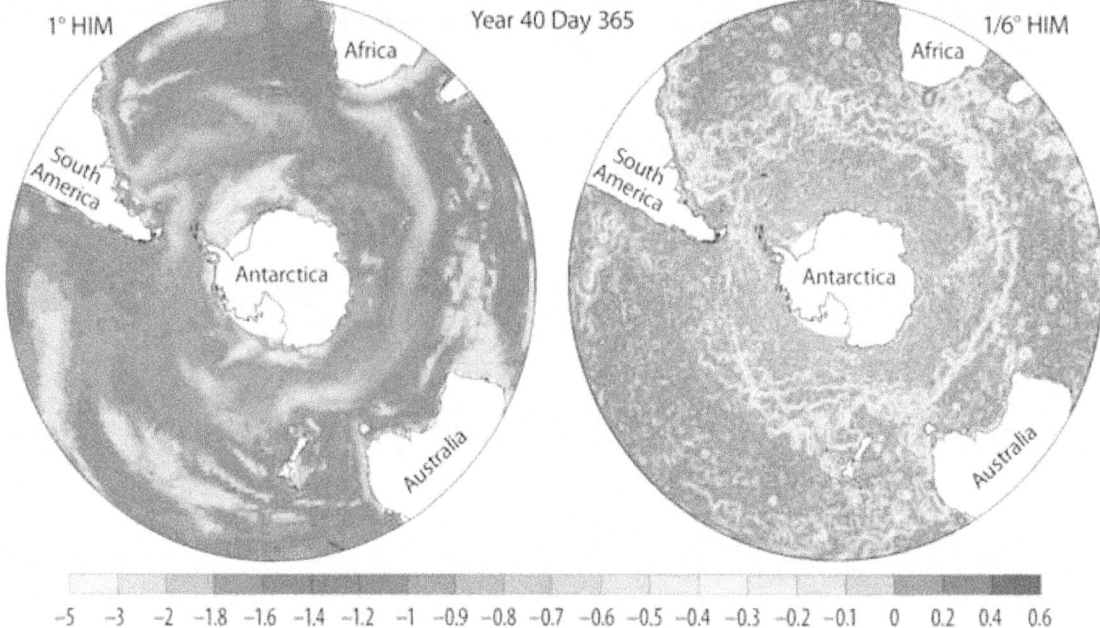

Ocean Surface Speed in NOAA/GFDL Southern Ocean Simulations

Log$_{10}$ of Magnitude of Velocity Averaged over Top 100 m in m s^{-1}

Figure 6.1. Surface-Current Speed in Two Simulations of the Southern Ocean in Low- and High-Resolution Ocean Models.

[From Fig. 6 in R. Hallberg and A. Gnanadesikan 2006: The role of eddies in determining the structure and response of the wind-driven Southern Hemisphere overturning: Results from the modeling eddies in the Southern Ocean (MESO) project. *J. Physical Oceanography*, **36**, 2232–2252. Reproduced by permission of the American Meteorological Society (AMS).]

Global mesoscale eddy-resolving ocean models are beginning to be examined in various modeling centers in the United States and around the world, even though exploiting such models will require substantial increases in computational resources. Challenges that may arise when these models are integrated for long time periods include maintaining realistically small amounts of mixing across constant-density surfaces in the more turbulent flows to avoid distortion of much slower thermohaline circulations.

As noted in Chapter 5, models provide estimates of the climate system's centennial-scale variability that underlies attribution studies of climatic trends. Seeing if eddy-resolving OGCMs increase the variability level on long time scales in climate models will be of great interest.

6.1.2 Cloud-Resolved Atmospheric Models

As atmospheric models attain higher resolution and more detailed representation of physical processes, short-range weather prediction and longer-range climate prediction become more synergistic (Phillips et al. 2004). This is particularly evident in "cloud-resolving models" (CRMs) with spatial resolutions of less than a few kilometers. CRMs can explicitly simulate atmospheric systems that exist on scales much smaller than the grid resolution of conventional atmospheric general circulation models (AGCMs) (Randall et al. 2003; Khairoutdinov, Randall, and DeMott 2005). These systems include mesoscale organizations in squall lines, deep updrafts and downdrafts, and cirrus anvils. CRMs also allow calculation of cloud properties and amounts based on more realistic small-scale structure in the flow field. The desired result is not only better simulations of regional climates, especially in the tropics, but also more reliable estimates of cloud feedbacks and climate sensitivity.

CRMs are variations of models designed for mesoscale storm and cumulus convection simulations. At CRM grid scales, hydrostatic balance is no longer universally valid. CRMs are therefore formulated with nonhydrostatic equations in which vertical accelerations are calculated explicitly (Tripoli 1992).

Like AGCMs, CRMs must employ empirical parameterizations to calculate the impact of subgrid scale processes, but CRMs explicitly represent a larger portion of the size spectrum of meteorological systems, so the parameterizations' impact on large-scale circulation and climate may be less severe. Most important, cumulus parameterizations for deep tropical convection are not needed in CRMs. CRMs can accommodate more realistic microphysical processes, including those by which aerosols nucleate cloud drops, allowing more convincing treatment of aerosol and cloud interactions involved in indirect aerosol radiative forcing.

However, shallow nonprecipitating convection (which produces fair-weather cumulus clouds) is dominated by flows on scales less than 1 km and will probably still require subgrid-scale parameterization in foreseeable global CRMs. Cloud feedbacks in regions of shallow convection are an important source of disparity in climate sensitivity in CMIP3 models (Bony et al. 2006). Furthermore, most cloud microphysical processes take place on CRM subgrid scales and so must be parameterized. Thus, uncertainty in cloud feedbacks will not disappear when global CRMs begin to play a role in climate assessments, but modelers hope that uncertainty will be reduced substantially.

Global models with CRM resolution have been attempted to date only at the Japanese Earth Simulator, but, with continued increase in computer power, global CRMs are expected to become centrally important in climate (as well as weather) research. Nevertheless, as noted above, major uncertainties in cloud microphysics will remain, especially in the prediction of ice-particle concentrations, fall speed of cloud particles, hydrometeorological spectra evolution, and entrainment rates into convective plumes (Cotton 2003). At CRM resolutions, more sophisticated algorithms of radiative-transfer calculation than those in current GCMs may be required because the plane parallel assumption for convergence of radiant energy may not be valid. Validation of CRMs probably will continue to take place in regional models and short-range forecasts, followed by their incorporation into global models.

Several observational programs such as the DOE Atmospheric Radiation Measurement

(ARM) Program have collected data essential to evaluate CRMs (M.H. Zhang et al. 2001; Tao et al. 2004). Results from such programs will facilitate improvement of CRM subgrid-scale physics. Extensive parameter-sensitivity tests with global models will still be needed to reduce uncertainties in microphysics and the treatment of shallow convection for climate sensitivity and regional climate-change simulation.

6.2 BIOGEOCHEMISTRY AND CLIMATE MODELS

6.2.1 Carbon Cycle

The physical climate system and biogeochemical processes are tightly coupled. Changes in climate affect the exchange of atmospheric CO_2 between land surface and ocean, and changes in CO_2 fluxes affect Earth's radiative forcing and thus the physical climate system. Some recently developed atmosphere-ocean general circulation models (AOGCMs) include the carbon cycle and have confirmed the potential for strong feedback between it and global climate (Cox et al. 2000; Friedlingstein et al. 2001; Govindasamy et al. 2005). The next generation of AOGCMs may include the carbon cycle as well as interactive atmospheric aerosols and chemistry. Models that include the carbon cycle are able to predict time-evolving atmospheric CO_2 concentrations using, as input, anthropogenic emissions rather than assumed concentrations.

Simulation of the global carbon cycle must account for the processes shown in Fig. 6.2. As the figure shows, the present-day global carbon cycle is not in equilibrium because of fossil-fuel burning and other anthropogenic carbon emissions. These carbon sources must, of course, be included in models of climate change. Such a calculation is not easy because human-induced changes to the carbon cycle are small compared to large natural fluxes, as shown in the figure. In addition, although the globally and annually averaged carbon reservoirs and fluxes shown in the figure are consistent with estimates from a variety of sources, substantial uncertainties are attached to the numbers (e.g., often a factor >2 uncertainty for fluxes; see Prentice et al. 2001). Additional uncertainty applies to regional, seasonal, and interannual variations in the carbon cycle.

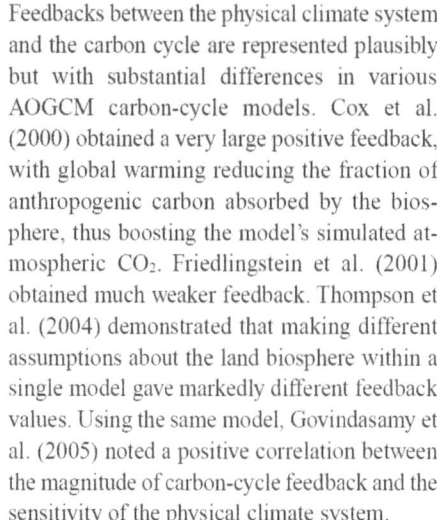

Feedbacks between the physical climate system and the carbon cycle are represented plausibly but with substantial differences in various AOGCM carbon-cycle models. Cox et al. (2000) obtained a very large positive feedback, with global warming reducing the fraction of anthropogenic carbon absorbed by the biosphere, thus boosting the model's simulated atmospheric CO_2. Friedlingstein et al. (2001) obtained much weaker feedback. Thompson et al. (2004) demonstrated that making different assumptions about the land biosphere within a single model gave markedly different feedback values. Using the same model, Govindasamy et al. (2005) noted a positive correlation between the magnitude of carbon-cycle feedback and the sensitivity of the physical climate system.

A recent study examined carbon-cycle feedbacks in 11 coupled AOGCM carbon-cycle models using the same forcing (Friedlingstein et al. 2006). The models unanimously agreed that global warming will reduce the fraction of anthropogenic carbon absorbed by the biosphere—a positive feedback—but the magnitude of this feedback varied widely among models (Fig. 6.3). When models included an interactive carbon cycle, predictions of the additional global warming due to carbon-cycle feedback ranged between 0.1 and 1.5°C. Eight models attributed most of the feedback to the land biosphere, while three attributed it to the ocean.

These results demonstrate the large sensitivity of climate model output to assumptions about carbon-cycle processes. Future carbon-cycle models, coupled to physical climate models and constrained by new global remote-sensing datasets and in situ measurements, may allow more definitive projection of CO_2 concentrations in the atmosphere for given emission scenarios. CCSP SAP 2.2 contains more information on the carbon cycle and climate change (CCSP 2007).

6.2.2 Other Biogeochemical Issues

Methane ($CH4$) is a potent greenhouse gas whose atmospheric concentration is controlled by its emission rate and the atmosphere's oxidative capacity (especially hydroxyl radical concentration). Methane concentrations are now much higher than in preindustrial times but have

Global Carbon Cycle as Seen by an AOGCM

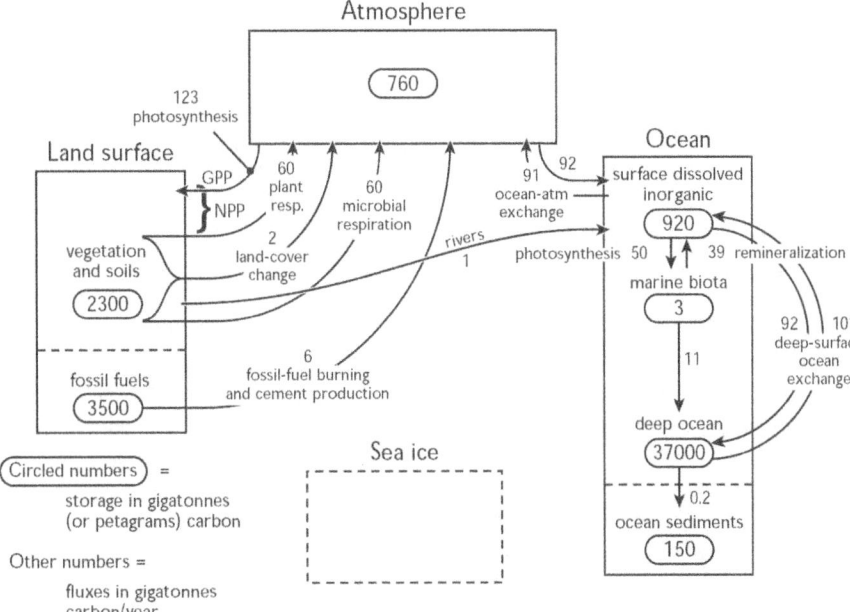

Figure 6.2. Global Carbon Cycle from the Point of View of Existing Physical Climate System Models (Coupled AOGCMs).

The four boxes represent atmosphere, land surface, ocean, and sea ice—major components of AOGCMs. Earth system models will evolve from AOGCMs by incorporating relevant biogeochemical cycles into the four-box framework (with sea ice not acting as a carbon reservoir). Numbers shown are average values for the 1990s. Small (≤1 PgC/year) fluxes such as those involving methane are not shown, except for burial of 0.2 PgC/year in ocean-bottom sediments, assuming a 50-50 split between plant and microbial respiration.

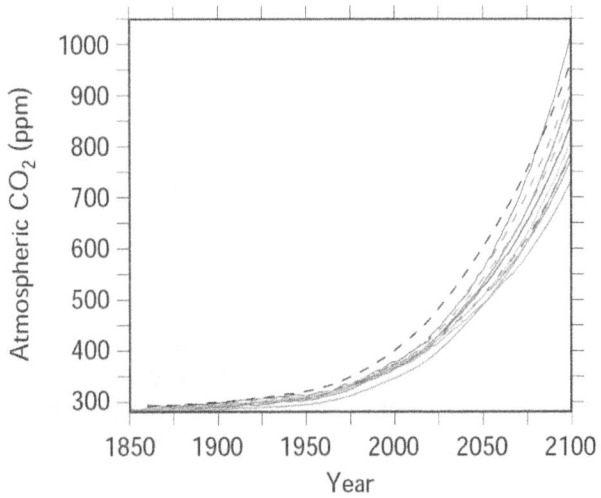

Figure 6.3. Time Series of Atmospheric CO₂ from 11 Different AOGCM Carbon-Cycle Models.

[From Fig. 1(a) of P. Friedlingstein et al. 2006: Climate-carbon cycle feedback analysis: Results from the C4MIP model intercomparison. *J. Climate*, **19**, 3337–3353. Reproduced by permission of the American Meteorological Society (AMS).]

not increased in the past decade, for reasons that continue to be debated. Whether or not this trend carries into the future has substantial implications for radiative forcing. To resolve this question, AOGCMs would need to include atmospheric chemistry models incorporating a number of different trace gases and reaction rates.

Another emerging issue is the interactive evolution of climate with the storage of water and carbon by plants. To address this process, dynamic vegetation models (in which plant growth is calculated rather than specified a priori) are under development at modeling centers in the United States and elsewhere. This inclusion of a wider range of processes poses challenges [e.g., it amplifies errors in rainfall prediction (Bonan and Levis 2006)]. In addition, ecosystems fertilized with CO₂ are limited by the availability of nutrients such as nitrogen and phosphorous that are important to the carbon cycle (Field, Jackson, and Mooney 1995; Schimel 1998; Nadelhoffer et al. 1999; Shaw et al. 2002; Hungate et al. 2003). Future climate-carbon models probably will need to include these nutrients. The few models that do so now show less plant growth in response to increasing atmospheric CO_2 (Cramer et al. 2001; Oren et al. 2001; Nowak, Ellsworth, and Smith 2004). Incorporation of other known limiting factors such as acclimation of soil microbiology to higher temperatures (Kirschbaum 2000; Tjoelker, Oleksyn, and Reich 2001) will be important in developing comprehensive Earth system models. Aerosol modeling also will be a central element in future models (this subject will be covered by CCSP SAP 2.3, whose estimated publication date is June 2008).

Often, climate-carbon simulations include natural ecosystems but do not include the effects of human land-cover and land-management changes (e.g., deforestation and reforestation). Land-cover change often is accounted for simply by prescribing estimates for the historical period (e.g., Houghton 2003) and for future scenarios from the IPCC Special Report on Emissions Scenarios (IPCC 2000). These estimates do not include practices such as crop irrigation and fertilization. Many models with "dynamic vegetation" do not actually simulate crops; they only allow natural vegetation to grow. Deforestation, land cultivation, and related human activities probably will be included in at least some future AOGCMs, enabling more complete assessment of total anthropogenic effects on the global climate and environment (Ramankutty et al. 2002; Root and Schneider 1993).

6.2.3 Ocean Biogeochemistry

Climate change impacts on the marine environment—including changes in the ocean's biota and carbon content due to modified ocean temperature, salinity, and circulation patterns—must be accounted for, along with terrestrial biogeochemistry, in a complete Earth system model. Implementation of ocean biogeochemistry processes into AOGCMs is under way to improve simulation of the ocean carbon cycle under various scenarios [e.g., "CCSM Biogeochemistry Working Group Meeting Report," March 2006 (www.ccsm.ucar.edu/

working_groups/Biogeo/reports/060328_BGC WGrpt.pdf); GFDL's Earth system model (gfdl.noaa.gov/~jpd/ esmdt.html); Doney et al. 2004]. One issue receiving particular attention in recent years is that ocean productivity may be increased through iron fertilization via dust particles, potentially reducing atmospheric CO_2 (Martin 1991). This effect is being assessed by both observational programs (e.g., Bishop, Davis, and Sherman 2002) and climate-carbon models (Jickells et al. 2005).

An important challenge to these efforts is the complexity of ocean ecosystems. Adding to this complexity are organisms that fix nitrogen and denitrify, calcify, or silicify; accounting for each adds parameterizations and variables to the system (Hood et al. 2006). Biological models need to be sufficiently complex to capture the observed variability on various time scales, since this variability provides essential tests for the models. As in many aspects of climate modeling, however, complexity that outgrows the ability to constrain models with available data should be avoided (Hood et al. 2006).

Modeling groups have undertaken systematic comparison of different models in the Ocean Carbon Cycle Model Intercomparison Project (OCMIP) under the auspices of the International Geosphere-Biosphere Programme. OCMIP's most recent phase involved 13 groups—including several from the United States—implementing a common biological model in their different OGCMs (Najjar et al. 2007). The common biological model includes five prognostic variables: inorganic phosphate (PO_4^{2-}), dissolved organic phosphorus (DOP), dissolved oxygen (O_2), dissolved inorganic carbon ($CO_2 + HCO_3^- + CO_3^{2-}$), and total alkalinity (the system's acid- and base-buffering capacity). Model intercomparison revealed significant differences in simulated biogeochemical fluxes and reservoirs. A biogeochemistry model's realism was found to be tied closely to the dynamics of the simulation's ocean circulation. Just as for land vegetation modeling, a serious challenge to climate models is presented by the quality of the physical climate simulation required for realistic biogeochemical modeling.

Example Applications of Climate Model Results

CHAPTER 7

In this chapter we present several cases where climate model simulation results were used for studies involving actual and potential end-user applications. With the increased availability of climate model simulation output through the CMIP3 multimodel archive, impacts and applications users are rapidly applying the model results for their needs. Just as quickly, the breadth and diversity of applications will continue to grow in the future as climate statistics are no longer considered stationary. The examples discussed in this chapter are meant for illustration and do not constitute a complete accounting of all published instances of applications from model results. The influence of climate, and therefore climate change, on different natural and societal systems is quite varied. Some impacts of climate change result primarily from changes in mean conditions. Other impacts are sensitive to climate variability—the sequence, frequency, and intensity of specific weather events. Note that the climate simulations described below are not offering predictions of 21st Century climate but simply projections of possible climate scenarios. Prediction requires knowing in advance how climatic forcings, including those produced by humans, would change in the future. SAP 3.2 examines climate projections by CMIP3 models in greater detail.

7.1 APPLYING MODEL RESULTS TO IMPACTS

As shown in previous chapters, climate models give approximate renditions of real climate. Consequently, applications of climate model results to impact studies require consideration of several limitations that characterize model output. In principle, using the direct output of climate models is desirable because these results represent a physically consistent picture of future climate, including changes in climate variability and the occurrence of such various weather phenomena as extreme events. In practice, this is rarely done for applications like those presented below because of simulation biases and the coarse spatial resolution of typical global simulations. Although the use of climate projections for impacts is beyond the scope of this report, aspects of the methodology for using the projections are based on the models' abilities to simulate observed climate. Employing coarse-resolution global model output for regional and local impact studies requires two additional steps—downscaling, as discussed in Chapter 3, and bias removal, or the adjustment of future projections for known systematic model errors, described in Chapters 2 and 5.

7.1.1 Downscaling

Downscaling is required because of the limitations of coarse spatial resolution in the global models. In mountainous terrain, a set of model values for a single grid box will represent conditions at the mean elevation level of that grid box. In reality, however, conditions at mountaintop and valley locations will be much different. Such processes as local snowpack accumulation and melting cannot be studied accurately with direct model output. Resolution also limits the accuracy of representation of small-scale processes. A prominent example is precipitation. The occurrence of heavy downpours is an important climate feature for certain impacts, but these events are often localized on a scale smaller than a grid box. In many actual situations, an area the size of a grid box may experience flooding rains at some points while others receive no rain at all. As a result, grid-box precipitation tends to be more frequent, and the largest values typically are smaller than those observed at the local scale. Chapter 3 covered both dynamical downscaling with nested regional models and statistical downscaling methods that include diverse techniques such as weather generators, transfer functions, and weather typing.

7.1.2 Bias Removal

A simple approach developed for bias removal during the early days of climate change assessments and still widely used today is sometimes dubbed the "delta" method. Climate model output is used to determine future change in climate with respect to the model's present-day climate, typically a difference for temperature and a percentage change for precipitation. Then, these changes are applied to observed historical climate data for input to an impacts model. The delta method assumes that future model biases for both mean and variability will be the same as those in present-day simulations. One highly questionable consequence of this assumption is that the future frequency and magnitude of extreme weather events are the same relative to the mean climate of the future as they are in present-day climate. Other bias-removal methods have been developed, but none are nearly so widespread, or they are versions of the delta method.

7.2 CALIFORNIA CLIMATE CHANGE ASSESSMENTS

One of the most comprehensive uses of climate model simulation output for applications is overseen by the California Climate Change Center. The center was established by a state agency, the California Energy Commission (CEC), through its Public Interest Energy Research program (CEC 2006). The center wanted to determine possible impacts of climate change on California and utilized the CMIP3 model simulation database as its starting point for climate change projections.

To generate future California scenarios, researchers selected three climate models from the CMIP3 multimodel archive: the National Center for Atmospheric Research–U.S. Department of Energy PCM, the NOAA GFDL CM2.1, and the Hadley Centre HadCM3 (Hayhoe et al. 2004; Cayan et al. 2006). The models were chosen in large part because of their ability to simulate both large-scale global climate features and California's multiple climatic regions when simulations of the 20th Century were compared with high-resolution observations. Of particular importance was the correct simulation of the state's precipitation climatology, with a pronounced wet season from November to March, during which nearly all annual precipitation falls. Further, these three models offered a range of sensitivities, with transient climate responses of 1.3 K for PCM, 1.5 K for CM2.1, and 2.0 K for HadCM3. Following model selection, projections from three scenarios with low, medium, and high future greenhouse gas emissions were chosen to span the range of possible future California climate states in the 21st Century. The California scenarios employed a statistical downscaling technique that, used observationally, derived probability density functions for surface temperature and precipitation to produce corrected model-simulated distribution functions (Cayan et al. 2006). Corrections were then applied to future scenario simulation results. Once the scenarios were generated, they were used to quantify possible climate change impacts on public health, water resources, agriculture, forests, and coastal regions (CEC 2006).

7.3 DRYLAND CROP YIELDS

The effects of weather and climate on crops are complex. Despite the fact that many details of weather interactions with plant physiology are poorly understood, numerous realistic crop-growth simulation models have been developed. Current-generation crop models typically step through the growth process with daily frequency and use a number of meteorological variables as input, typically maximum and minimum temperature, precipitation, solar radiation, and potential evapotranspiration. A key characteristic of these models is that they have been developed for application at a single location and have been validated based on point data, including meteorological inputs. Thus, their use in assessing climate change impacts on crop yields confronts a mismatch between the spatially averaged climate model grid-box data and the point data expected by crop models. Also, biases in climate model data can have unknown effects on crop model results because the dependence of crop yields on meteorological variables is highly nonlinear. The typical application study circumvents these difficulties by avoiding the direct use of climate model output.

The delta method continues to be a common approach in contemporary crop studies. In the U.S. National Assessment of the Consequences of Climate Variability and Change, monthly changes (model future – model control) were applied to observed data, and a weather generator was used to produce daily weather data for input to impacts models. For example, Winkler et al. (2002) found a longer growing season and greater seasonal heat accumulation in fruit-growing regions of the Great Lakes but uncertainty about future susceptibility to freezes. Olesen et al. (2007) investigated the potential impacts of climate change on several European crops. Crop models were driven by direct output of regional climate models and also baseline (present-day) observed daily climate data adjusted by GCM changes using the delta method. Thomson et al. (2005) adjusted current daily climate data with monthly change values derived from GCM projections (Smith et al. 2005) and then used them as input to models to study future yields of dryland crops in the United States. National yield changes were found to be up to ± 25%, depending on the climate scenario. These

applications of the delta method produce daily climate unchanged in many respects from present-day observed data. The number of precipitation days and the time between them remains the same. Also, relative changes in intensity are the same for light and heavy days. Likewise, the length of extended periods of extreme heat and cold and the intensity of such extremes with respect to the new climate mean do not change.

In a recent study, Zhang (2005) used statistical downscaling to estimate Oklahoma wheat yields for a future simulation from HadCM3. In this study, mean monthly changes of the means and variances of temperature and precipitation between the HadCM3 control and future simulations were used to adjust the parameters of a weather generator model. Weather generator parameters include mean precipitation, precipitation variance, the probability of a wet day following a wet day, the probability of a dry day following a wet day, mean temperature, and temperature variance. The observed data were used to determine a relationship between the wet-wet and wet-dry day probabilities and total monthly precipitation. This relationship was used to assign future values of those probabilities based on the GCM-simulated precipitation changes. With the new set of parameters, the weather generator simulated multiple years of daily weather variables for input to the yield model. This approach is logical and consistent and produces different variability characteristics depending on whether future climate is wetter or drier than the present, unlike the simple delta method applied to daily climate data. However, these changes are assumed to be similar to what occurs in the present-day climate between wet and dry periods. Thus, more subtle climate model–simulated changes that might affect yields (e.g., a change to longer wet and dry spells without a change in total precipitation) are not transmitted.

7.4 SMALL WATERSHED FLOODING

This application faces many of the same issues as applying model output to estimate changes in dryland crop yields. For example, models used for simulating runoff in small watersheds have been validated using point station data. In addition, runoff is a highly nonlinear function

of precipitation, and flooding occurrence is particularly sensitive to the exact frequency and amount of precipitation for the most extreme events. As noted in the *"Extreme Events"* section of Chapter 5, climate models often underestimate the magnitude of extremes. Again, the delta method is frequently applied to estimate the changes in flooding that may result from global climate change. Recently, Cameron (2006) determined percentage changes in precipitation from climate model simulations and applied them to a stochastic rainfall model to produce precipitation time series for input to a hydrologic model. Flood magnitudes were estimated for return periods of 10 to 200 years and for several climate changes scenarios. In most cases, flood flows increased, but one scenario produced a decrease.

Dibike and Coulibaly (2005) applied two statistical downscaling techniques to an analysis of flow on a small watershed in northern Quebec. One technique used the model of Wilby, Dawson, and Barrow (2002) to identify a set of large-scale variables (i.e., pressure, flow, temperature, and humidity) related to surface temperature and precipitation in the watershed. The resulting statistical relationships were applied to the output of a Canadian GCM climate change simulation to generate future surface temperature and precipitation time series. The second technique used a weather generator requiring various statistical parameters, estimated by comparing surface temperature and precipitation data between GCM control and future scenario simulations. The fundamental difference between these two statistical downscaling techniques is that the Wilby, Dawson, and Barrow (2002) model uses a more complete set of atmospheric data from the GCM output data while the weather generator uses only surface temperature and precipitation. The resulting time series from both methods provided input for a hydrologic model. In both cases, peak flows are higher in the spring and lower in the early summer in future warmer climates, reflecting changes in snowmelt timing. A major difference is that the Wilby, Dawson, and Barrow (2002) model produces a trend of increasing daily precipitation not seen in the weather generator data, resulting in larger spring increases in peak flow.

7.5 URBAN HEAT WAVES

This estimation of changes in heat-wave frequency and intensity can be accomplished using only near-surface temperature. Because heat waves are large-scale phenomena and near-surface temperature is rather highly correlated over the scales of GCM grid-boxes, downscaling is not usually required for their analysis. Biases, while remaining an issue, can be accounted for by using percentile-based definitions of heat waves. Meehl and Tebaldi (2004) used output from the PCM for 2080 to 2099 to calculate percentile-based measures of extreme heat; they found that heat waves will increase in intensity, frequency, and duration. If mortality estimates are desired, then biases are an issue because existing models (Kalkstein and Greene 1997) used location-specific absolute magnitudes of temperature to estimate mortality.

7.6 WATER RESOURCES IN THE WESTERN UNITED STATES

The possibility that climate change may adversely affect limited water resources in the mostly arid and semiarid western United States poses a threat to the prosperity of that region. A group of university and government scientists, under the auspices of the U.S Department of Energy–sponsored Accelerated Climate Prediction Initiative Pilot Project, conducted a coordinated set of studies that represented an end-to-end assessment of this issue (Barnett et al. 2004). This project is noteworthy because of close coordination between production of GCM simulations and the needs of impacts modeling. It also is a good example of more-sophisticated downscaling approaches.

A suite of carefully selected PCM climate simulations was executed (Dai et al. 2004; Pierce 2004) and then used to drive a regional climate model to provide higher-resolution data (Leung et al. 2004), both for direct assessment of effects on water resources and for use in impacts models. A careful statistical downscaling approach (Wood et al. 2004) also was used to produce an alternate dataset for input to impacts models. Using the observationally based 1/8° latitude-by-longitude resolution gridded dataset developed by Maurer et al. (2002), an empirical mapping function was developed to relate quantiles of the simulated monthly temperature and precipitation frequency distributions from control runs to the observed climatological monthly distributions at the GCM grid scale. This empirical mapping was then applied to simulated future monthly temperature and precipitation data and spatially disaggregated to the 1/8° resolution grid through a procedure that added small-scale structure. Daily time series of future climate on the 1/8° grid subsequently were produced by randomly sampling from historical data and adding in the changes resulting from the empirical mapping and disaggregation.

The daily time series were used in a set of studies to assess water resource impacts (Stewart, Cayan, and Dettinger 2004; Payne et al. 2004; VanRheenen et al. 2004; Christensen et al. 2004). The studies, which assumed the IPCC business-as-usual emissions scenario for the climate change GCM simulation, indicate that warmer temperatures will melt the snowpack about a month earlier throughout western North America by the end of the 21st Century. The shift in snowmelt will decrease flows and increase competition for water during the summer in the Columbia River Basin (Payne et al. 2004). In the Sacramento River and San Joaquin River basins, the average April 1 snowpack is projected to decrease by half. In the Colorado River basin, a decrease in total precipitation would mean that total system demand would exceed river inflows.

REFERENCES

AchutaRao, K.M., et al., 2006: Variability of ocean heat uptake: Reconciling observations and models. *J. Geophysical Research*, **111**, doi:10.1029/2005JC003136.

AchutaRao, K.M., and K.R. Sperber, 2006: ENSO simulation in coupled ocean-atmosphere models: Are the current models better? *Climate Dynamics*, doi:10.1007/s00382-006-0119-7.

AchutaRao, K., and K.R. Sperber, 2002: Simulation of the El Niño Southern Oscillation: Results from the Coupled Model Intercomparison Project. *Climate Dynamics*, **19**, 191–209.

Ackerman, A.S., et al., 2004: The impact of humidity above stratiform clouds on indirect aerosol climate forcing. *Nature*, **432**, 1014–1017.

Ackerman, T. P., and G.M. Stokes, 2003: The Atmospheric Radiation Measurement Program. *Physics Today*, **56**(1), 38–44.

Adams, R.M., B.A. McCarl, and O. Mearns, 2003: The effects of spatial scale climate scenarios on economic assessments: An example from U.S. agriculture. *Climate Change*, **60**, 131–148.

Albrecht, B.A., 1989: Aerosols, cloud microphysics, and fractional cloudiness. *Science*, **245**, 1227–1230.

Alexander, M., et al., 2006: Extratropical atmosphere-ocean variability in CCSM3. *J. Climate*, **19**, 2496–2525.

Alley, R.B., et al., 2005: Ice-sheet and sea-level changes. *Science*, **310**, 456–460.

Ammann, C.M., et al., 2003: A monthly and latitudinally varying volcanic forcing dataset in simulations of 20th Century climate. *Geophysical Research Letters*, **30**(12), 1657.

Anderson, B.T., et al., 2001: Model dynamics of summertime low-level jets over northwestern Mexico. J. *Geophysical Research*, **106**(D4), 3401–3413.

Anderson, C.J., R.W. Arritt, and J.S. Kain, 2007: An alternative mass flux profile in the Kain-Fritsch convective parameterization and its effects in seasonal precipitation. J. *Hydrometeorology*, **8**(5), 1128–1140, doi:10.1175/JHM624.1.

Anderson, C.J., et al., 2003: Hydrologic processes in regional climate model simulations of the central United States flood of June–July 1993. *J. Hydrometeorology*, 4, 584–598.

Anderson, T.L., et al., 2003: Climate forcing by aerosols—a hazy picture. *Science*, **300**, 1103–1104.

Andersson, L., et al., 2006: Impact of climate change and development scenarios on flow patterns in the Okavango River. *J. Hydrology*, **331**, 43–57.

Annamalai, H., K. Hamilton, and K.R. Sperber, 2007: South Asian summer monsoon and its relationship with ENSO in the IPCC AR4 simulations. *J. Climate*, **20**(6), 1071–1092.

Annan, J.D., and J.C. Hargreaves, 2006: Using multiple observationally based constraints to estimate climate sensitivity. *Geophysical Research Letters*, **33**(6), doi:10.1029/2005GL025259.

Annan, J.D., et al., 2005: Efficiently constraining climate sensitivity with paleoclimate simulations. *SOLA*, **1**, 181–184.

Antic, S., et al., 2006: Testing the downscaling ability of a one-way nested regional climate model in regions of complex topography. *Climate Dynamics*, **26**, 305–325.

Antic, S., et al., 2004: Testing the downscaling ability of a one-way nested regional climate model in regions of complex topography. *Climate Dynamics*, **23**(5), 473–493.

Antonov, J.I., S. Levitus, and T.P. Boyer, 2005: Thermosteric sea level rise, 1955–2003. *Geophysical Research Letters*, **32**, L12602.

Arakawa, A., and W.H. Schubert, 1974: Interaction of a cumulus cloud ensemble with the large-scale environment. Part I. *J. Atmospheric Sciences*, **31**, 674–701.

Arblaster, J.M., and G.A. Meehl, 2006: Contribution of various external forcings to trends in the Southern Annular Mode. *J. Climate*, **19**, 2896–2905.

Arendt, A.A., et al., 2002: Rapid wastage of Alaska glaciers and their contribution to rising sea level. *Science*, **297**, 382–386.

Arora, V.K., and G.J. Boer, 2003: A representation of variable root distribution in dynamic vegetation models. *Earth Interactions*, 7, 1–19.

Arzel, O., T. Fichefet, and H. Goosse, 2006: Sea ice evolution over the 20th and 21st centuries as simulated by current AOGCMs. *Ocean Modelling*, **12**, 401–415.

Aurela, M., T. Laurila, and J.P. Tuovinen, 2004: The timing of snow melt controls the annual CO_2 balance in a subarctic fen. *Geophysical Research Letters*, **31**, L16119.

Avissar, R., and R.A. Pielke, 1989: A parameterization of heterogeneous land-surface for atmospheric numerical models and its impact on regional meteorology. *Monthly Weather Review*, **117**, 2113–2136.

Baldocchi, D., and P. Harley, 1995: Scaling carbon dioxide and water vapour exchange from leaf to canopy in a deciduous forest. I. Leaf model parameterization. *Plant, Cell, Environment*, **18**, 1146–1156.

Bard, E., B. Hamelin, and R.G. Fairbanks, 1990: U-Th ages obtained by mass spectrometry in corals from Barbados: Sea level during the past 130,000 years. *Nature*, **346**, 456–458.

Barnett, T.P., et al., 2005: Penetration of human-induced warming into the world's oceans. *Science*, **309**, 284.

Barnett, T., et al., 2004: The effects of climate change on water resources in the West: Introduction and overview. *Climatic Change*, **62**, 1–11.

Beckman, A., and R. Doscher, 1997: A method for improved representation of dense water spreading over topography in geopotential-coordinate models. *J. Physical Oceanography*, **27**, 581–591.

Bell, J.L., L.C. Sloan, and M.A. Snyder, 2004: Regional changes in extreme climatic events: A future climate scenario. *J. Climate*, **17**(1), 81–87.

Bellouin, N., et al., 2005: Global estimate of aerosol direct radiative forcing from satellite measurements. *Nature*, **438**, 1138–1141.

Bengtsson, L., K.I. Hodges, and M. Esch, 2007: Tropical cyclones in a T159 resolution global climate model: Comparison with observations and re-analyses. *Tellus A*, **59**(4), 396–416, doi:10.1111/j.1600-0870.2007.00236.x.

Beringer, J., et al., 2001: The representation of arctic soils in the land surface model: The importance of mosses. *J. Climate*, **14**, 3324–3335.

Betts, A.K., and J.H. Ball, 1997: Albedo over the boreal forest. *J. Geophysical Research–Atmosphere*, **102**, 28901–28909.

Biasutti, M., and A. Giannini, 2006: Robust Sahel drying in response to late 20th Century forcings. *Geophysical Research Letters*, **33**, L11706.

Bierbaum, R.M., et al., 2003: *Estimating Climate Sensitivity: Report of a Workshop*. National Academy of Sciences, Washington, D.C. (books.nap.edu/catalog/10787.html).

Bishop, J.K.B., R.E. Davis, and J.T. Sherman, 2002: Robotic observations of dust storm enhancement of carbon biomass in the North Pacific. *Science*, **298**, 817–821.

Bitz, C.M., and W.H. Lipscomb, 1999: An energy-conserving thermodynamic model of sea ice. *J. Geophysical Research*, **104**, 15669–12677.

Bjerknes, J., 1969: Atmospheric teleconnections from the equatorial Pacific. *Monthly Weather Review*, **97**, 163–172.

Bleck, R., 2002: An oceanic general circulation model framed in hybrid isopycnic-Cartesian coordinates. *Ocean Modelling*, **4**, 55–88.

Bleck, R., et al., 1992: Salinity-driven thermocline transients in a wind- and thermohaline-forced isopycnic coordinate model of the North Atlantic. *J. Physical Oceanography*, **22**(12), 1485–1505.

Boer, G., J. Stowasser, and K. Hamilton, 2007: Inferring climate sensitivity from volcanic events. *Climate Dynamics*, **28**(5), 481–502.

Boisserie, M., et al., 2006: Evaluation of soil moisture in the Florida State University climate model: National Center for Atmospheric Research community land model (FSU-CLM) using two reanalyses (R2 and ERA40) and in situ observations. *J. Geophysical Research*, **111**(D8), Art. No. D08103.

Bonan, G. B., and S. Levis, 2006: Evaluating aspects of the community land and atmosphere models (CLM3 and CAM3) using a dynamic global vegetation model. J. *Climate*, **19**, 2290–2301.

Bonan, G.B., 1995: Sensitivity of a GCM simulation to inclusion of inland water surfaces. *J. Climate*, **8**, 2691–2704.

Bonan, G.B., D. Pollard, and S.L. Thompson, 1992: Effects of boreal forest vegetation on global climate. *Nature*, **359**, 716–718.

Bony, S., et al., 2006: How well do we understand and evaluate climate change feedback processes? *J. Climate*, **19**(15), 3445–3482.

Bony, S., and J.-L. Dufresne, 2005: Marine boundary layer clouds at the heart of tropical cloud feedback uncertainties in climate models. *Geophysical Research Letters*, **32**, L20806.

Bony, S., et al., 2004: On dynamic and thermodynamic components of cloud changes. *Climate Dynamics*, **22**, 71–86.

Bony, S., and K.A. Emanuel, 2001: A parameterization of the cloudiness associated with cumulus convection: Evaluation using TOGA COARE data. *J. Atmospheric Sciences*, **58**, 3158–3183.

Boone, A., et al., 2000: The influence of the inclusion of soil freezing on simulations by a soil-vegetation-atmosphere transfer scheme. *J. Applied Meteorology*, **39**, 1544–1569.

Boyle, J.S., et al., 2008: Climate model forecast experiments for TOGA-COARE. *Monthly Weather Review*, in press.

Boyle, J.S., 1993: Sensitivity of dynamical quantities to horizontal resolution for a climate simulation using the ECMWF (cycle 33) model. *J. Climate*, **6**, 796–815.

Bougeault, P., et al., 2001: The MAP special observing period. *Bulletin American Meteorological Society*, **82**(3), 433–462.

Braconnot, P., et al., 2007a: Results of PMIP2 coupled simulations of the mid-Holocene and Last Glacial Maximum. Part 1: Experiments and large-scale features. *Climate Past*, **3**(2), 261–277.

Braconnot, P., et al., 2007b: Results of PMIP2 coupled simulations of the Mid-Holocene and Last Glacial Maximum. Part 2: Feedbacks with emphasis on the location of the ITCZ and mid- and high-latitudes heat budget. *Climate Past* **3**(2), 279–296.

Braithwaite, R.J., and S.C.B. Raper, 2002: Glaciers and their contribution to sea level change. *Physics Chemistry Earth*, **27**, 1445–1454.

Brankovic, T., and D. Gregory, 2001: Impact of horizontal resolution on seasonal integrations. *Climate Dynamics*, **18**, 123–143.

Breugem, W.-P., W. Hazeleger, and R.J. Haarsma, 2006: Multi-model study of tropical Atlantic variability and change. *Geophysical Research Letters*, **33**, L23706.

Briegleb, B.P., et al., 2002: *Description of the Community Climate System Model Version 2 Sea Ice Model*, 60 pp. (www.ccsm.ucar.edu/models/ccsm2.0/csim/).

Brinkop, S., and E. Roeckner, 1995: Sensitivity of a general circulation model to parameterizations of cloud-turbulence interactions in the atmospheric boundary layer. *Tellus*, **47A**, 197–220.

Brown, T.J., B.L. Hall, and A.L. Westerling, 2004: The impact of Twenty-First Century climate change on wildland fire danger in the western United States: An applications perspective. *Climatic Change*, **62**, 365–388.

Bryan, F.O., et al., 2006: Changes in ocean ventilation during the 21st Century in the CCSM3. *Ocean Modelling*, **15**, 141–156.

Bryan, K., 1969a: A numerical method for the study of the circulation of the world ocean. *J. Computational Physics*, **4**, 347–376.

Bryan, K., 1969b: Climate and the ocean circulation. III: The ocean model. *Monthly Weather Review*, **97**, 806–824.

Bryan, K., and M.D. Cox, 1967: A numerical investigation of the oceanic general circulation. *Tellus*, **19**, 54–80.

Bryden, H.L., and S. Imawaki, 2001: Ocean heat transport, 455–474 in Siedler, G., J. Church, and J. Gould, eds. *Ocean Circulation and Climate: Observing and Modelling the Global Ocean*, Academic Press, San Diego, p.109.

Burke, E.J., S.J. Brown, and N. Christidis, 2006: Modeling the recent evolution of global drought and projections for the 21st Century with the Hadley Centre climate model. *J. Hydrometeorology*, **7**, 1113–1125.

Byerle, L.A., and J. Paegle, 2003: Modulation of the Great Plains low-level jet and moisture transports by orography and large-scale circulations. *J. Geophysical Research*, **108**(D16), Art. No. 8611.

Cai, W.J., P. Whetton, and D.J. Karoly, 2003: The response of the Antarctic Oscillation to increasing and stabilized atmospheric CO_2. *J. Climate*, **16**, 1525–1538.

California Energy Commission (CEC), 2006: *Our Changing Climate: Assessing the Risks to California, A Summary Report from the California Climate Change Center*, 16 pp. (www.climatechange.ca.gov/biennial_reports/2006report/index.html).

Callaghan, T.V., et al., 2004: Responses to projected changes in climate and UV-B at the species level. *Ambio*, **33**, 418–435.

Cameron, D., 2006: An application of the UKCIP02 climate change scenarios to flood estimation by continuous simulation for a gauged catchment in the northeast of Scotland, U.K. (with uncertainty). *J. Hydrology*, **328**, 212–226.

Camp, C.D., and K.K. Tung, 2007: Surface warming by the solar cycle as revealed by the composite mean difference projection. *Geophysical Research Letters*, **34**, L14703, doi:10.1029/2007GL030207.

Cane, M.A., et al., 1997: Twentieth Century sea surface temperature trends. *Science*, **275**, 957–960.

Capotondi, A., A. Wittenberg, and S. Masina, 2006: Spatial and temporal structure of ENSO in 20th Century coupled simulations. *Ocean Modelling*, **15**, 274–298.

Carril, A.F., C.G. Menéndez, and A. Navarra, 2005: Climate response associated with the Southern Annular Mode in the surroundings of Antarctic Peninsula: A multimodel ensemble analysis. *Geophysical Research Letters*, **32**, L16713.

Cassano, J.J., et al., 2007: Predicted changes in synoptic forcing of net precipitation in large Arctic river basins during the 21st Century. *J. Geophysical Research*, **112**, G04S49, doi:10.1029/2006JG000332.

Cassano, J.J., P. Uotilla, and A.H. Lynch, 2006: Changes in synoptic weather patterns in the polar regions in the 20th and 21st centuries. Part 1: Arctic. *International J. Climatology*, **26**(9), 1181–1199.

Cayan, D., et al., 2006: *Climate Scenarios from California: A Report from California Climate Change Center*, White Paper, 52 pp. (www.energy.ca.gov/2005publications/CEC-500-2005-203/CEC-500-2005-203.SF.PDF).

Cazenave, A., and R.S. Nerem, 2004: Present-day sea level change: Observations and causes. *Reviews Geophysics*, **42**(3), RG3001.

CCSP, 2007: *The First State of the Carbon Cycle Report (SOCCR): The North American Carbon Budget and Implications for the Global Carbon Cycle*, ed. A.W. King, et al. Synthesis and Assessment Product 2.2, U.S. Climate Change Science Program and Subcommittee on Global Change Research, Washington, D.C. (www.climate-science.gov/Library/sap/sap2-2/final-report/default.htm).

CCSP, 2006: *Temperature Trends in the Lower Atmosphere: Steps for Understanding and Reconciling Differences*, ed. T.R. Karl et al. Synthesis and Assessment Product 1.1, U.S. Climate Change Science Program and Subcommittee on Global Change Research, Washington, D.C. (www.climatescience.gov/Library/sap/sap1-1/finalreport/default.htm).

Cess, R.D., et al., 1990: Intercomparison and interpretation of climate feedback processes in 19 atmospheric general circulation models. *J. Geophysical Research*, **95**(D10), 16601–16615.

Cess, R.D., and G.L. Potter, 1988: Exploratory studies of cloud radiative forcing with a general circulation model. *Tellus*, **39A**, 460–473.

Charney, J.G., 1979: *Carbon Dioxide and Climate: A Scientific Assessment*. National Academy of Sciences, Washington, D.C., 22 pp.

Chen, F., and J. Dudhia, 2001: Coupling an advanced land surface–hydrology model with the Penn State–NCAR MM5 modeling system. Part I: Model implementation and sensitivity. *Monthly Weather Review*, **129**, 569–585.

Cheng, A., and K.-M. Xu, 2006: Simulation of shallow cumuli and their transition to deep convective clouds by cloud-resolving models with different third-order turbulence closures. *Quarterly J. Royal Meteorological Society*, **132**, 359–382.

Cheng, Y., V.M. Canuto, and A.M. Howard, 2002: An improved model for the turbulent BBL. *J. Atmospheric Sciences*, **59**, 1550–1565.

Chiacchio, M., J. Francis, and P. Stackhouse, Jr., 2002: Evaluation of methods to estimate the surface downwelling longwave flux during Arctic winter. *J. Applied Meteorology*, **41**, 306–318.

Choi, H.I., P. Kumar, and X.-Z. Liang, 2007: Three-dimensional averaged soil-moisture transport model with a scalable parameterization of subgrid topographic variability. *Water Resources Research*, **43**, WO4414.

Christensen, J.H., et al., 2007: *Regional Climate Projections in Climate Change 2007: The Physical Science Basis. Contribution of Working Group 1 to the Fourth Assessment Report of the Intergovernmental panel on Climate Change*, ed. S. Solomon et al. Cambridge University Press, Cambridge, U.K., and New York, N.Y.

Christensen, J.H., T. Carter, and F. Giorgi, 2002: PRUDENCE employs new methods to assess European climate change. *Eos*, **83**, 147.

Christensen, N.S., et al., 2004: The effects of climate change on the hydrology and water resources of the Colorado River basin. *Climatic Change*, **62**, 337–363.

Christensen, O.B., 1999: Relaxation of soil variables in a regional climate model. *Tellus*, **51A**, 674–685.

Christidis, N., et al., 2005: Detection of changes in temperature extremes during the second half of the 20th Century. *Geophysical Research Letters*, **32**, L20716.

Christy, J.R., et al., 2007: Tropospheric temperature change since 1979 from tropical radiosonde and satellite measurements. *J. Geophysical Research*, **112**, D06102, doi:10.1029/2005JD006881.

Christy, J. R., and R. W. Spencer, 2005: Correcting temperature data sets. *Science*, **310**, 972.

Chung, C.E., et al., 2005: Global anthropogenic aerosol direct forcing derived from satellite and ground-based observations. *J. Geophysical Research*, **110**, D24207.

Church, J.A., and N.J. White, 2006: A 20th Century acceleration in global sea-level rise. *Geophysical Research Letters*, **33**, L01602.

Church, J.A., N.J. White, and M. Arblaster, 2005: Significant decadal-scale impact of volcanic eruptions on sea level and ocean heat content. *Nature*, **438**(7064), 74–77, doi:10.1038/nature04237.

Claussen, M., et al., 2002: Earth system models of intermediate complexity: Closing the gap in the spectrum of climate system models. *Climate Dynamics*, **18**, 579–586.

Clement, A.C., and B. Soden, 2005: The sensitivity of the tropical-mean radiation budget. *J. Climate*, **18**, 3189–3203.

Clement, A.C., et al., 1996: An ocean dynamical thermostat. *J. Climate*, **9**, 2190–2196.

Clough, S.A., M.J. Iacono, and J.-L. Moncet, 1992: Line-by-line calculations of atmospheric fluxes and cooling rates: Application to water vapor. *J. Geophysical Research*, **97**, 15761–15785.

Collins, M., et al., 2006: Towards quantifying uncertainty in transient climate change. *Climate Dynamics*, **27**(2–3), 127–147.

Collins, W.D., et al., 2006(a): The Community Climate System Model Version 3: CCSM3. *J. Climate*, **19**(11), 2122–2143.

Collins, W.D., et al., 2006(b): Radiative forcing by well-mixed greenhouse gases: Estimates from climate models in the Intergovernmental Panel on Climate Change (IPCC) Fourth Assessment Report (AR4). *J. Geophysical Research*, **111**, D14317.

Collins, W.D., et al., 2004: Description of the NCAR Community Atmospheric Model (CAM3). *NCAR Technical Note*, National Center for Atmospheric Research, NCAR/TN-464+STR, 226 pp.

Colman, R., 2003: A comparison: Climate feedbacks in general circulation models. *Climate Dynamics*, **20**, 865–873.

Cook, K.H., and E.K. Vizy, 2006: Coupled model simulations of the West African monsoon system: 20th Century simulations and 21st Century predictions. *J. Climate*, **19**, 3681–3703.

Coppola, E., and F. Giorgi, 2005: Climate change in tropical regions from high-resolution time-slice AGCM experiments. *Quarterly J. Royal Meteorological Society*, **131**, 3123–3145.

Cotton, W.R., 2003: Cloud modeling from days of EML to the present—Have we made progress? 95–106 in *AMS Meteorological Monographs—Symposium on Cloud Systems, Hurricanes, and TRMM*.

Covey, C., et al., 2003: An overview of results from the Coupled Model Intercomparison Project, *Global Planetary Change*, **37**(1–2), 103–133.

Covey, C., et al., 2000: The seasonal cycle in coupled ocean-atmosphere general circulation models. *Climate Dynamics*, **16**, 775–787.

Covey, C., L.C. Sloan, and M.I. Hoffert, 1996: Paleoclimate data constraints on climate sensitivity: The paleocalibration method. *Climatic Change*, **32**, 165–184.

Cox, P.M., et al., 2000: Acceleration of global warming due to carbon-cycle feedbacks in a coupled model. *Nature*, **408**, 184–187.

Cramer, W., et al., 2001: Global response of terrestrial ecosystem and function to CO_2 and climate change: Results from six dynamic global vegetation models. *Global Change Biology*, **7**, 357–373.

Crane, R.G., and B.C. Hewitson, 1998: Doubled CO_2 precipitation changes for the Susquehanna basin: Down-scaling from the GENESIS general circulation model. *International J. Climatology*, **18**, 65–76.

Crucifix, M., 2006: Does the last glacial maximum constrain climate sensitivity? *Geophysical Research Letters*, **33**, L18701, doi:10.1029/2006GL027137.

Crucifix, M., et al., 2006: Second phase of Paleoclimate Modelling Intercomparison Project. EOS, *Transactions American Geophysical Union*, **86**(28), 264.

Cubasch, U., et al., 2001: Projections of future climate change, 525–582 in *Climate Change 2001: The Scientific Basis*, ed. J.T. Houghton, et al. Cambridge University Press, Cambridge, U.K.

Cunningham, S.A., et al., 2003: Transport and variability of the Antarctic Circumpolar Current in Drake Passage. *J. Geophysical Research*, 108, doi:10.1029/2001JC001147.

Curry, J.A., and A.H. Lynch, 2002: Comparing Arctic regional climate models. *Eos*, **83**, 87.

Curry, J.A., et al., 1996: Overview of Arctic cloud and radiation characteristics. *J. Climate*, **9**, 1731–1764.

Curry, J.A., J. Schramm, and E.E. Ebert, 1995: Sea ice-albedo climate feedback mechanism. *J. Climate*, **8**, 240–247.

Cusack, S., J.M. Edwards, and R. Kershaw, 1999: Estimating subgrid variance of saturation and its parameterization for use in a GCM cloud scheme. *Quarterly J. Royal Meteorological Society*, **125**, 3057–3076.

Dai, A., 2006: Precipitation characteristics in eighteen coupled climate models. *J. Climate*, **19**, 4605–4630.

Dai, A., et al., 2004: The ACPI climate change simulations. *Climatic Change*, **62**, 29–43.

Danabasoglu, G., et al., 2006: Diurnal coupling in the tropical oceans of CCSM3. *J. Climate*, **19**, 2347–2365, doi:10.1175/JCLI3739.1.

Davies, H.C., and R.E. Turner, 1977: Updating prediction models by dynamical relaxation: An examination of the technique. *International J. Climatology*, **103**, 225–245.

Davies, H.C., 1976: Lateral boundary formulation for multilevel prediction models. *Quarterly J. Royal Meteorological Society*, **102**, 405–418.

DeCaria, A.J., et al., 2005: Lightning-generated NOx and its impact on tropospheric ozone production: A three-dimensional modeling study of a stratosphere–troposphere experiment: Radiation, aerosols, and ozone (STERAO-A) thunderstorm. *J. Geophysical Research*, **110**, D14303, doi:10.1029/2004JD00556.

de Elía, R., R. Laprise, and B. Denis, 2002: Forcasting skill limits of nested, limited-area models: A perfect-model approach. *Monthly Weather Review*, **130**, 2006–2023.

Del Genio, A.D., et al., 2005: Cumulus microphysics and climate sensitivity. *J. Climate*, **18**, 2376–2387, doi:10.1175/JCLI3413.1.

Del Genio, A.D., A. Wolf, and M.-S. Yao, 2005: Evaluation of regional cloud feedbacks using single-column models. *J. Geophysical Research*, **110**, D15S13, doi:10.1029/2004JD005011.

Del Genio, A.D., et al., 1996: A prognostic cloud water parameterization for global climate models. *J. Climate*, **9**, 270–304.

Del Genio, A.D., and M.-S. Yao, 1993: Efficient cumulus parameterization for long-term climate studies: The GISS scheme. *The Representation of Cumulus Convection in Numeric Models. Meteorological Monograph*, No. 46, American Meteorological Society, 181–184.

Delworth, T.L., et al., 2006: GFDL's CM2 global coupled climate models. Part 1: Formulation and simulation characteristics. *J. Climate*, **19**, 643–674.

Denis, B., R. Laprise, and D. Caya, 2003: Sensitivity of a regional climate model to the resolution of the lateral boundary conditions. *Climate Dynamics*, **20**, 107–126.

Denis, B., et al., 2002: Downscaling ability of one-way, nested regional climate models: The big-brother experiments. *Climate Dynamics*, **18**, 627–646.

Déqué, M., et al., 2005: Global high resolution versus Limited Area Model climate change projections over Europe: Quantifying confidence level from PRUDENCE results. *Climate Dynamics*, **25**, 653–670.

Déqué, M., and J.P. Piedelievre, 1995: High-resolution climate simulation over Europe. *Climate Dynamics*, **11**, 321–339.

Dethloff, K., et al., 2006: A dynamical link between the Arctic and the global climate system. *Geophysical Research Letters*, **33**, L03703, doi:10.1029/2005GL025245.

Dettinger, M.D., et al., 2004: Simulated hydrologic responses to climate variations and change in the Merced, Carson, and American River basins, Sierra Nevada, California, 1900–2099. *Climatic Change*, **62**, 283–317.

Dibike, Y.B., and P. Coulibaly, 2005: Hydrologic impact of climate change in the Saguenay watershed: Comparison of downscaling methods and hydrologic models. *J. Hydrology*, **307**, 145–163.

Dickinson, R.E., A. Henderson-Sellers, and P.J. Kennedy, 1993: Biosphere-Atmosphere Transfer Scheme (BATS) version 1e as coupled to the NCAR Community Climate Model. *NCAR Technical Note*, NCAR/TN-387+STR, National Center for Atmospheric Research, Boulder, Colo., 72 pp. [Available from NCAR; P.O. Box 3000; Boulder, CO 80305]

Diffenbaugh, N.S., M.A. Snyder, and L.C. Sloan, 2004: Could CO_2-induced land-cover feedbacks alter near-shore upwelling regimes? *Proceedings National Academy Sciences*, **101**, 27–32.

Dimitrijevic, M., and R. Laprise, 2005: Validation of the nesting technique in a regional climate model and sensitivity tests to the resolution of the lateral boundary conditions during summer. *Climate Dynamics*, **25**, 555–580.

Dimotakis, P.E., 2005: Turbulent mixing. *Annual Review Fluid Mechanics*, **37**, 329–356.

Doney, S.C., et al., 2004: *Ocean Carbon and Climate Change (OCCC): An Implementation Strategy for U.S. Ocean Carbon Cycle Science*. University Corporation for Atmospheric Research, Boulder, Colo., 108 pp.

Douglass, D.H., and R.S. Knox, 2005: Climate forcing by the volcanic eruption of Mount Pinatubo. *Geophysical Research Letters*, **32**, L05710, doi:10.1029/2004GL022119.

Ducharne, A., et al., 2003: Development of a high resolution runoff routing model, calibration, and application to assess runoff from the LMD GCM. *J. Hydrology*, **280**, 207–228.

Duffy, P.B., et al., 2003: High-resolution simulations of global climate. Part 1: Present climate. *Climate Dynamics*, **21**, 371–390.

Dutay, J.-C., et al., 2002: Evaluation of ocean model ventilation with CFC-11: Comparison of 13 global ocean models. *Ocean Modelling*, **4**, 89–120.

Duynkerke, P.G., and S.R. de Roode, 2001: Surface energy balance and turbulence characteristics observed at the SHEBA ice camp during FIRE III. *J. Geophysical Research*, **106**, 15313–15322.

Easterling, D.R., 2002: Recent changes in frost days and the frost-free season in the United States. *Bulletin American Meteorological Society*, **83**, 1327–1332.

Ebert, E.E., J.L. Schramm, and J.A. Curry, 1995: Disposition of solar radiation in sea ice and the upper ocean. *J. Geophysical Research*, **100**, 15965–15975.

Ebert, E.E., and J.A. Curry, 1993: An intermediate one-dimensional thermodynamic sea ice model for investigating ice-atmosphere interactions. *J. Geophysical Research*, **98**, 10085–10109.

Ellingson, R., and Y. Fouquart, 1991: The intercomparison of radiation codes in climate models: An overview. *J. Geophysical Research*, **96**(D5), 8925–8927.

Emanuel, K.A., 1994: *Atmospheric Convection*. Oxford University Press, Oxford, U.K., 580 pp.

Emanuel, K.A., 1991: A scheme for representing cumulus convection in large-scale models. *J. Atmospheric Sciences*, **48**, 2313–2335.

Emori, S., and S.J. Brown, 2005: Dynamic and thermodynamic changes in mean and extreme precipitation under changed climate. *Geophysical Research Letters*, **32**, L17706.

Emori, S., et al., 2005: Validation, parameterization dependence, and future projection of daily precipitation simulated with a high-resolution atmospheric GCM. *Geophysical Research Letters*, **32**(6), L06708.

Enfield, D.B., A.M. Mestas-Nuñez, and P.J. Trimble, 2001: The Atlantic Multidecadal Oscillation and its relationship to rainfall and river flows in the continental U.S., *Geophysical Research Letters*, **28**, 2077–2080.

Essery, R., and J. Pomeroy, 2004: Implications of spatial distributions of snow mass and melt rate for snow-cover depletion: Theoretical considerations. *Annals Glaciology*, **38**, 261–265.

Fedorov, A., and S.G. Philander, 2000: Is El Niño changing? *Science*, **288**, 1997–2002.

Field, C., R. Jackson, and H. Mooney, 1995: Stomatal responses to increased CO_2: Implications from the plant to the global scale. *Plant, Cell, Environment*, **18**, 1214–1225.

Forest, C.E., P.H. Stone, and A.P. Sokolov, 2006: Estimated PDFs of climate system properties including natural and anthropogenic forcings. *Geophysical Research Letters*, **33**, L01705, doi:10.1029/2005GL023977.

Forster, P., et al., 2007: Changes in Atmospheric Constituents and in Radiative Forcing. In *Climate Change 2007: The Physical Science Basis. Contribution of Working Group 1 to the Fourth Assessment of the Intergovernmental Panel on Climate Change*, ed. S. Solomon et al. Cambridge University Press, Cambridge, U.K., and New York, N.Y.

Foukal, P., et al., 2006: Variations in solar luminosity and their effect on the Earth's climate. *Nature*, **443**, 161–166.

Fox-Rabinovitz, M.S., et al., 2006: Variable resolution general circulation models: Stretched-grid model intercomparison project (SGMIP). *J. Geophysical Research*, **111**(D16), Art. No. D16104.

Fox-Rabinovitz, M.S., L.L. Takacs, and R.C. Govindaraju, 2002: A variable-resolution stretched-grid general circulation model and data assimilation system with multiple areas of interest: Studying the anomalous regional climate events of 1998. *J. Geophysical Research*, **107**(D24), Art. No. 4768.

Fox-Rabinovitz, M.S., et al., 2001: A variable-resolution stretched-grid general circulation model: Regional climate simulation. *Monthly Weather Review*, **129**, 453–469.

Fox-Rabinovitz, M.S., and R.S. Lindzen, 1993: Numerical experiments on consistent horizontal and vertical resolution for atmospheric models and observing systems. *Monthly Weather Review*, **121**, 264–271.

Frame, D.J., et al., 2005: Constraining climate forecasts: The role of prior assumptions. *Geophysical Research Letters*, **32**, L09702, doi:10.1029/2004GL022241.

Frei, C., et al., 2006: Future change of precipitation extremes in Europe: An intercomparison of scenarios from regional climate models. *J. Geophysical Research*, **111**(D6), Art. No. D06105.

Frich, P., et al., 2002: Observed coherent changes in climatic extremes during the second half of the Twentieth Century. *Climate Research*, **19**, 193–212.

Fridlind, A.M., et al., 2004: Evidence for the predominance of mid-tropospheric aerosols as subtropical anvil cloud nuclei. *Science*, **304**(5671), 718–722.

Friedlingstein, P., et al., 2006: Climate-carbon cycle feedback analysis: Results from the C4MIP model intercomparison. *J. Climate*, **19**, 3337–3353.

Friedlingstein, P., et al., 2001: Positive feedback between future climate change and the carbon cycle. *Geophysical Research Letters*, **28**(8), 1543–1546.

Fröhlich, C., 2006: Solar irradiance variability since 1978. *Space Science Reviews*, **125**, 53–65, doi:10.1007/s11214-006-9046-5.

Fu, C.B., et al., 2005: Regional climate model intercomparison project for Asia. *Bulletin American Meteorological Society*, **86**, 257–266.

Fyfe, J.C., G.J. Boer, and G.M. Flato, 1999: The Arctic and Antarctic oscillations and their projected changes under global warming. *Geophysical Research Letters*, **11**, 1601–1604.

Ganachaud, A., and C. Wunsch, 2000: Improved estimates of global ocean circulation, heat transport and mixing from hydrographic data. *Nature*, **408**, 453–457.

Gao, S., L. Ran, and X. Li, 2006: Impacts of ice microphysics on rainfall and thermodynamic processes in the tropical deep convective regime: A 2D cloud-resolving modeling study. *Monthly Weather Review*, **134**(10), 3015–3024.

GARP, 1975: *The Physical Basis of Climate Modeling*, Global Atmospheric Research Program (GARP), Publication Series #16, April 1975.

Gates, W.L., et al., 1999: An Overview of the Atmospheric Model Intercomparison Project (AMIP). *Bulletin American Meteorological Society*, **80**(1), 29–55, doi:10.1175/1520-0477(1999)080<0029:AOOTRO.2.0.CO;2.

Gates, W.L., 1992: AMIP: The Atmospheric Model Intercomparison Project. *Bulletin American Meteorological Society*, **73**, 1962–1970.

Gent, P.R., et al., 2006: Ocean chlorofluorocarbon and heat uptake during the 20th Century in the CCSM3. *J. Climate*, **19**, 2366–2381.

Gent, P., and J.C. McWilliams, 1990: Isopycnal mixing in ocean circulation models. *J. Physical Oceanography*, **20**, 150–155.

GFDL Global Atmospheric Model Development Team (GAMDT), 2004: The new GFDL global atmosphere and land model AM2/LM2: Evaluation with prescribed SST conditions. *J. Climate*, **17**(24), 4641–4673.

Ghan, S. J., et al., 2000: An intercomparison of single column model simulations of summertime midlatitude continental convection. *J. Geophysical Research*, **105**, 2091–2124.

Gillett, N.P., and D.W.J. Thompson, 2003: Simulation of recent Southern Hemisphere climate change. *Science*, **302**, 273–275, doi:10.1126/science.1087440.

Giorgi, F., and L.O. Mearns, 2003: Probability of regional climate change based on the reliability ensemble averaging (REA) method. *Geophysical Research Letters*, **30**, Art. No. 1629.

Giorgi, F., et al., 2001: Regional climate change information—Evaluation and projections. Chapter 10, 583–638, in *Climate Change 2001: The Scientific Basis*, ed. J.T. Houghton, et al. Cambridge University Press, Cambridge, U.K.

Giorgi, F., and X. Bi, 2000: A study of internal variability of a regional climate model. *J. Geophysical Research*, **105**, 29503–29521.

Giorgi, F., and L.O. Mearns, 1999: Introduction to special section: Regional climate modeling revisited. *J. Geophysical Research*, **104**(D6), 6335–6352.

Giorgi, F., et al., 1996: A regional model study of the importance of local versus remote controls of the 1988 drought and the 1993 flood over the central United States. *J. Climate*, **9**, 1150–1162.

Giorgi, F., M.R. Marinucci, and G.T. Bates, 1993: Development of a second-generation regional climate model (RegCM2). Part II: Convective processes and assimilation of lateral boundary conditions. *Monthly Weather Review*, **121**, 2814–2832.

Giorgi, F., and L.O. Mearns, 1991: Approaches to the simulation of regional climate change—A review. *Reviews Geophysics*, **29**, 191–216.

Gleckler, P.J., K.E. Taylor, and C. Doutriaux, 2008: Performance metrics for climate models, *J. Geophysical Research*, **113**, D06104, doi:10.1029/2007JD008972.

Gleckler, P.J., K.R. Sperber, and K. AchutaRao, 2006: Annual cycle of global ocean heat content: Observed and simulated. *J. Geophysical Research*, **111**, C06008.

Gnanadesikan, A., et al., 2006: GFDL's CM2 global coupled climate models. Part II: The baseline ocean simulation. *J. Climate*, **19**, 675–697.

Gorham, E., 1991: Northern peatlands: Role in the carbon cycle and probable responses to climatic warming. *Ecological Applications*, **1**, 182–195.

Govindasamy, B., et al., 2005: Increase of carbon cycle feedback with climate sensitivity: Results from a coupled climate and carbon cycle model. *Tellus*, **57B**, 153–163.

Grabowski, W.W., 2003: MJO-like coherent structures: Sensitivity simulations using the cloud-resolving convection parameterization (CRCP). *J. Atmospheric Sciences*, **60**, 847–864.

Grabowski, W.W., and M.W. Moncrieff, 2001: Large-scale organization of tropical convection in two-dimensional explicit numerical simulations. *International J. Climatology*, **127**, 445–468.

Gregory, J.M., J.A. Lowe, and S.F.B. Tett, 2006: Simulated global-mean sea level changes over the last half-millennium. *J. Climate*, **19**, 4576–4592.

Gregory, J.M., et al., 2004: Simulated and observed decadal variability in ocean heat content. *Geophysical Research Letters*, **31**(15), L15312.

Gregory, J.M., 1999: Representation of the radiative effect of convective anvils. *Hadley Centre Technical Note 7*, Hadley Centre for Climate Prediction and Research, Met Office, Fitzroy Road, Exeter, EX1 3BP, U.K.

Gregory, J.M., and J.F.B. Mitchell, 1997: The climate response to CO_2 of the Hadley Centre coupled AOGCM with and without flux adjustment. *Geophysical Research Letters*, **24**, 1943–1946.

Gregory, D., and S. Allen, 1991: The Effect of Convective Scale Downdrafts upon NWP and Climate Simulations, 122-123 in *Ninth Conference on Numerical Weather Prediction*. American Meteorological Society, Denver, Colo.

Gregory, D., and P.R. Rowntree, 1990: A mass flux convection scheme with representation of cloud ensemble characteristics and stability dependent closure. *Monthly Weather Review*, **118**, 1483–1506.

Grell, G.A., et al., 2000a: Application of a multiscale, coupled MM5/chemistry model to the complex terrain of the VOTALP valley campaign. *Atmospheric Environment*, **34**, 1435–1453.

Grell, G.A., et al., 2000b: Nonhydrostatic climate simulations of precipitation over complex terrain. *J. Geophysical Research–Atmospheres*, **105** (D24), 29595–29608.

Grell, G.A., H. Dudhia, and D. S. Stanfler, 1994: A description of the fifth generation Penn State–NCAR Mesoscale Model (MM5). *NCAR Technical Note*. NCAR/TN-3981STR, National Center for Atmospheric Research, Boulder, Colo., 122 pp. [Available from NCAR; P.O. Box 3000; Boulder, CO 80305]

Grell, G.A., 1993: Prognostic evaluation of assumptions used by cumulus parameterizations. *Monthly Weather Review*, **121**, 764–787.

Griffies, S.M., et al., 2005: Formulation of an ocean model for global climate simulations. *Ocean Science*, **1**, 45–79.

Griffies, S.M., et al., 2001: Tracer conservation with an explicit free surface method for z-coordinate ocean models. *Monthly Weather Review*, **129**, 1081–1098.

Griffies, S.M., 1998: The Gent-McWilliams skew-flux. *J. Physical Oceanography*, **28**, 831–841.

Gritsun, A., and G. Branstator, 2007: Climate response using a three-dimensional operator based on the fluctuation–dissipation theorem. *J. Atmospheric Sciences*, **64**(7), 2558–2575

Gu, L., et al., 1999: Micrometeorology, biophysical exchanges, and NEE decomposition in a two-storey boreal forest—Development and test of an integrated model. *Agricultural Forest Meteorology*, **94**, 123–148.

Guilyardi, E., 2006: El Niño-mean state-seasonal cycle interactions in a multi–model ensemble. *Climate Dynamics*, **26**, 329–348.

Guo, Z., D.H. Bromwich, and J.J. Cassano, 2003: Evaluation of polar MM5 simulations of Antarctic atmospheric circulation. *Monthly Weather Review*, **131**, 384–411.

Gutowski, W.J., et al., 2007a: A possible constraint on regional precipitation intensity changes under global warming. *J. Hydrometeorology*, **8**, 1382–1396, doi: 10.1175/2007JHM817.1.

Gutowski, W.J., et al., 2007b: Influence of Arctic wetlands on Arctic atmospheric circulation. *J. Climate*, **20**, 4243–4254, doi:10.1175/JCL14243.1.

Gutowski, W.J., et al., 2003: Temporal-spatial scales of observed and simulated precipitation in central U.S. climate. *J. Climate*, **16**, 3841–3847.

Gutowski, W.J., et al., 2002: A Coupled Land–Atmosphere Simulation Program (CLASP). *J. Geophysical Research*, **107**(D16), 4283, doi:1029/2001JD000392.

Gutowski, W.J., Z. Ötles, and Y. Chen. 1998: Effect of ocean-surface heterogeneity on climate simulation. *Monthly Weather Review*, **126**, 1419–1429.

Hack, J.J., 1994: Parameterization of moist convection in the National Center for Atmospheric Research Community Climate Model (CCM2). *J. Geophysical Research*, **99**, 5551–5568.

Hagemann, S., and L. Dümenil, 1998: A parameterization of the lateral water flow for the global scale. *Climate Dynamics*, **14**, 17–31.

Hall, A., 2004: The role of surface albedo feedback in climate. *J. Climate*, **17**, 1550–1568.

Hallberg, R., and A. Gnanadesikan, 2006: The role of eddies in determining the structure and response of the wind-driven Southern Hemisphere overturning: Results from the Modeling Eddies in the Southern Ocean (MESO) project. *J. Physical Oceanography*, **36**, 2232–2252.

Haltiner, G.J., and R.T. Williams, 1980: *Numerical Prediction and Dynamic Meteorology*, Second Edition, John Wiley & Sons, New York, 477 pp.

Han, J., and J.O. Roads, 2004: U.S. climate sensitivity simulated with the NCEP regional spectral model. *Climate Change*, **62**, 115–154.

Hansen, J., et al., 2007: Climate simulations for 1880–2003 with GISS modelE. *Climate Dynamics*, **29**(7-8), 661–696 (arxiv.org/abs/physics/0610109).

Hansen, J., et al., 2006: Global temperature change. *Proceedings National Academy Sciences*, **103**, 14288–14293, doi:10.1073/pnas.0606291103.

Hansen, J., et al., 2005a: Earth's energy imbalance: Confirmation and implications. *Science*, **308**, 1431–1435.

Hansen, J., et al., 2005b: Efficacy climate forcings. *J. Geophysical Research*, **110**, D18104, doi:10.1029/2005JD005776.

Hansen, J., et al., 1996: Global surface air temperature in 1995: Return to pre-Pinatubo level. *Geophysical Research Letters*, **23**, 1665–1668, doi:10.1029/96GL01040.

Hansen, J., et al., 1993: How sensitive is the world's climate? *National Geographic Research Exploration*, **9**, 42–158.

Hansen, J., et al., 1985: Climate response times: Dependence on climate sensitivity and ocean mixing. *Science*, **229**, 857–859.

Hansen, J., et al., 1984: Climate Sensitivity: Analysis of Feedback Mechanisms, 130–163 in *Climate Processes and Climate Sensitivity*. Maurice Ewing Series, 5, ed. J.E. Hansen and T. Takahashi. American Geophysical Union, Washington, D.C.

Hansen, J., et al., 1981: Climate impact of increasing atmospheric carbon dioxide. *Science*, **213**, 957–966.

Harries, J.E., et al., 2001: Increases in greenhouse forcing inferred from the outgoing longwave radiation spectra of the Earth in 1970 and 1997. *Nature*, **410**, 355–357.

Hartmann, D.L., and K. Larson, 2002: An important constraint on tropical cloud–climate feedback. *Geophysical Research Letters*, **29**, 1951, doi:10.1029/2002GL015835.

Hartmann, D.L., and M.L. Michelsen, 2002: No evidence for iris. *Bulletin American Meteorological Society*, **83**(2), 249–254.

Hay, L.E., et al., 2006: One-way coupling of an atmospheric and a hydrologic model in Colorado. *J. Hydrometeorology*, **7**, 569–589.

Hayhoe, K., et al., 2004: Emissions pathways, climate change, and impacts on California. *Proceedings National Academy Sciences* **101**(34), 12422–12427.

Haylock, M.R., et al., 2006: Downscaling heavy precipitation over the U.K.: A comparison of dynamical and statistical methods and their future scenarios. *International J. Climatology*, **26**, 1397–1415.

Hegerl, G., et al. 2007: Understanding and attributing climate change. In Climate Change 2007: *The Physical Science Basis. Contribution of Working Group 1 to the Fourth Assessment of the Intergovernmental Panel on Climate Change*. Cambridge University Press, Cambridge, U.K., and New York, N.Y.

Hegerl, G.C., et al., 2006: Climate sensitivity constrained by temperature reconstructions over the past seven centuries. *Nature*, **440**, 1029–1032.

Hegerl, G.C., et al., 2004: Detectability of anthropogenic changes in annual temperature and precipitation extremes. *J. Climate*, **17**(19), 3683–3700.

Held, I.M., and B.J. Soden, 2006: Robust responses of the hydrological cycle to global warming. *J. Climate*, **19**, 5686–5699.

Held, I.M., et al., 2005: Simulation of Sahel drought in the 20th and 21st centuries. *Proceedings National Academy Sciences*, **102**, 17891–17896.

Held, I.M., and B.J. Soden, 2000: Water vapor feedback and global warming. *Annual Review of Energy Environment*, **25**, 441–475.

Held, I., and M.J. Suarez, 1994: A proposal for the intercomparison of the dynamical cores of atmospheric general circulation models. *Bulletin American Meteorological Society*, **75**(10), 1825–1830.

Held, I.M., R.S. Hemler, and V. Ramaswamy, 1993: Radiative-convective equilibrium with explicit two-dimensional moist convection. *J. Atmospheric Sciences*, **50**, 3909–3927.

Held, I.M., S.W. Lyons, and S. Nigam, 1989: Transients and the extratropical response to El Niño. *J. Atmospheric Sciences*, **46**(1), 163–174.

Helfand, H.M., and J.C. Labraga, 1988: Design of a non-singular level 2.5 second order closure model for the prediction of atmospheric turbulence. *J. Atmospheric Sciences*, **45**, 113–132.

Hellstrom, C., et al., 2001: Comparison climate change scenarios for Sweden based on statistical and dynamical downscaling of monthly precipitation. *Climate Research*, **19**, 45–55.

Henderson-Sellers, A., 2006: Improving land-surface parameterization schemes using stable water isotopes: Introducing the "iPILPS" initiative. *Global Planetary Change*, **51**, 3–24.

Henderson-Sellers, A., et al., 2003: Predicting land-surface climates—Better skill or moving targets? *Geophysical Research Letters*, **30**, 1777.

Henderson-Sellers, A., et al., 1995: The Project for Intercomparison of Land-Surface Parameterization Schemes (PILPS) —Phase 2 and Phase 3. *Bulletin American Meteorological Society*, **76**, 489–503.

Hewitson, B.C., and R.G. Crane, 1996: Climate downscaling: Techniques and application. *Climate Research*, **7**, 85–95.

Hewitt, C.D., and D.J. Griggs, 2004: Ensembles-based predictions: Climate changes and their impacts. *Eos*, **85**, 566.

Heyen, H., H. Fock, and W. Greve, 1998: Detecting relationships between the interannual variability in ecological time series and climate using a multivariate statistical approach—A case study on Helgoland Roads zooplankton. *Climate Research*, **10**, 179–191.

Hibler, W.D., 1979: A dynamic thermodynamic sea ice model. *J. Physical Oceanography*, **9**, 815–846.

Hines, K.M., et al., 1999: Surface energy balance of the NCEP MRF and NCEP-NCAR reanalysis in Antarctic latitudes during FROST. *Weather Forecasting*, **14**, 851–866.

Hirano, A., R. Welch, and H. Lang, 2003: Mapping from ASTER stereo image data: DEM validation and accuracy assessment. ISPRS *J. Photogrammetry Remote Sensing*, **57**, 356–370.

Hoerling, M., et al., 2006: Detection and attribution of 20th Century northern and southern African rainfall change. *J. Climate*, **19**, 3989–4008.

Hoffert, M.I., and C. Covey, 1992: Deriving global climate sensitivity from paleoclimate reconstructions. *Nature*, **360**, 573–576.

Holland, M.M., and M.N. Raphael, 2006: Twentieth Century simulation of the Southern Hemisphere climate in coupled models. Part II: Sea ice conditions and variability. *Climate Dynamics*, **26**, 229–245.

Holland, M.M., and C.M. Bitz, 2003: Polar amplification: Climate change in coupled models. *Climate Dynamics*, **21**, 221–232.

Holtslag, A.A.M., and B.A. Boville, 1993: Local versus nonlocal boundary-layer diffusion in a global climate model. *J. Climate*, **6**, 1825–1842.

Hong, S.-Y., and H.-M.H. Juang, 1998: Orography blending in the lateral boundary of a regional model. *Monthly Weather Review*, **126**, 1714–1718.

Hood, R.R., et al., 2006: Pelagic functional group modeling: Progress, challenges and prospects. *Deep-Sea Research II*, **53**, 459–512.

Hoogenboom, G., J.W. Jones, and K.J. Boote, 1992: Modeling growth, development, and yield of grain legumes using SOYGRO, PNUTGRO, and BEANGRO—A review. *Transactions ASAE*, **35**, 2043–2056.

Hope, P.K., N. Nicholls, and J.L. McGregor, 2004: The rainfall response to permanent inland water in Australia. *Australian Meteorological Magazine*, **53**, 251–262.

Horel, J.D., and J.M. Wallace, 1981: Planetary-scale atmospheric phenomena associated with the Southern Oscillation. *Monthly Weather Review*, **109**, 813–829.

Hori, M.E., and H. Ueda, 2006: Impact of global warming on the East Asian winter monsoon as revealed by nine coupled atmosphere–ocean GCMs. *Geophysical Research Letters*, **33**, doi:10.1029/2005GL024961.

Horowitz, L.W., et al., 2003: A global simulation of tropospheric ozone and related tracers: Description and evaluation of MOZART, version 2. *J. Geophysical Research–Atmospheres*, **108**(D24), 4784.

Houghton, R., 2003: Revised estimates of the annual net flux of carbon to the atmosphere from changes in land use and land management 1850–2000. *Tellus*, **55B**, 378–390.

Huang, B., P.H. Stone, and C. Hill, 2003: Sensitivities of deep-ocean heat uptake and heat content to surface fluxes and subgridscale parameters in an ocean general circulation model with idealized geometry. *J. Geophysical Research*, **108**(C1), 3015, doi:10.1029/2001JC001218.

Hungate, B.A., et al., 2003: Nitrogen and climate change. *Science*, **302**, 1512–1513.

Hunke, E.C., and Y. Zhang, 1999: A comparison of sea ice dynamics models at high resolution. *Monthly Weather Review*, **127**, 396–408.

Hunke, E.C., and J.K. Dukowicz, 1997: An elastic-viscous-plastic model for sea ice dynamics. *J. Physical Oceanography*, **27**, 1849–1867.

Hurrell, J.W., 1995: Decadal trends in the North Atlantic Oscillation and relationships to regional temperature and precipitation. *Science*, **269**, 676–679.

Iorio, J.P., et al., 2004: Effects of model resolution and subgrid-scale physics on the simulation of precipitation in the continental United States. *Climate Dynamics*, **23**, 243–258.

IPCC, 2007a: Couplings Between Changes in the *Climate System and Biogeochemistry. In Climate Change 2007: The Physical Science Basis. Contribution of Working Group 1 to the Fourth Assessment Report of the Intergovernmental Panel on Climate Change*, ed. S. Solomon et al. Cambridge University Press, Cambridge, U.K., and New York (www.ipcc.ch).

IPCC, 2007b: Summary for Policymakers. In *Climate Change 2007: The Physical Science Basis. Contribution of Working Group 1 to the Fourth Assessment Report of the Intergovernmental Panel on Climate Change*, ed. S. Solomon et al. Cambridge University Press, Cambridge, U.K., and New York (www.ipcc.ch).

IPCC, 2001: *Climate Change 2001: The Scientific Basis: Contribution of Working Group 1 to the Third Assessment Report of the Intergovernmental Panel on Climate Change*, ed. J.T. Houghton et al. Cambridge University Press, Cambridge, U.K., and New York.

IPCC, 2000: *Special Report on Emissions Scenarios*, ed. N. Nakicenovic and R. Swart (www.grida.no/climate/ipcc/emission/).

IPCC, 1990: *Climate Change: The IPCC Scientific Assessment*, ed. J.T. Houghton, G.J. Jenkins, and J.J. Ephraums. Cambridge University Press, Cambridge, U.K.

Irannejad, P., A. Henderson-Sellers, and S. Sharmeen, 2003: Importance of land-surface parameterization for latent heat simulation. *Geophysical Research Letters*, **30**, 1904.

Jakob, C., and G. Tselioudis, 2003: Objective identification of cloud regimes in the tropical western Pacific. *Geophysical Research Letters*, **30**, 2082.

Janowiak, J.E., 1988: An investigation of interannual rainfall variability in Africa. *J. Climate*, **1**, 240–255.

Jickells, T.D., et al., 2005: Global iron connections between desert dust, ocean biogeochemistry, and climate. *Science*, **308**, 67–71.

Jones, C.A., and J.R. Kiniry, 1986: *CERES-Maize: A Simulation Model of Maize Growth and Development*. Texas A&M University Press, College Station, Tex.

Jones, P.D., et al., 2006: Global and hemispheric temperature anomalies—land and marine instrumental records. Carbon Dioxide Information Analysis Center, Oak Ridge National Laboratory, U.S. DOE (cdiac.esd.ornl.gov/trends/temp/jonescru/jones.html).

Jones, P.D., et al., 1999: Surface air temperature and its changes over the past 150 years. *Reviews Geophysics*, **37**, 173–199.

Jones, P.D., T. Jónsson, and D. Wheeler, 1997: Extension to the North Atlantic Oscillation using early instrumental pressure observations from Gibraltar and south-west Iceland. *International J. Climatology*, **17**, 1433–1450.

Jones, R.G., J.M. Murphy, and M. Noguer, 1995: Simulation of climate change over Europe using a nested regional climate model. Part I: Assessment of control climate, including sensitivity to location of lateral boundaries. *International J. Climatology*, **121**, 1413–1449.

Joseph, R., and S. Nigam, 2006: ENSO evolution and teleconnections in IPCC's 20th Century climate simulations: Realistic representation? *J. Climate* **19**(17), 4360–4377.

Kain, J.S., and J.M. Fritsch, 1993: Convective Parameterization in Mesoscale Models: The Kain-Fritsch Scheme. In *The Representation of Cumulus Convection in Numerical Models, Meteorological Monograph*s. American Meteorological Society **46**, 165–170.

Kalkstein, L.S., and J.S. Greene, 1997: An evaluation of climate and mortality relationships in larger U.S. cities and the possible impacts of a climate change. *Environmental Health Perspectives*, **105**, 84–93.

Kalnay, E., et al., 1996: The NCEP/NCAR 40-year reanalysis project. *Bulletin American Meteorological Society*, **77**(3), 437–471.

Kanamitsu, M., et al., 2002: NCEP-DEOAMIP-II reanalysis (R-2). *Bulletin American Meteorological Society*, **83**, 1631–1643.

Kattenberg, A., et al., 1996: Climate Models Projections of Future Climate. Chapter 6, Climate Models–Projections of Future Climate, 285–358, in *Climate Change 1995–The Science. Climate Change*, ed. J.T. Houghton, et al. Cambridge University Press, Cambridge, U.K.

Kattsov, V., and E. Källén, 2005: Future Changes of Climate: Modelling and Scenarios for the *Arctic Region in Arctic Climate Impact Assessment (ACIA)*. Cambridge University Press, Cambridge, U.K., 1042 pp.

Kattsov, V.M., et al., 2000: Atmospheric Climate Models: Simulation of the Arctic Ocean Fresh Water Budget Components, 209–247 in *The Freshwater Budget of the Arctic Ocean*, ed. E.L. Lewis. Kluwer Academic Publishers, Dordrecht, The Netherlands.

Khain, A., and A. Pokrovsky, 2004: Simulation of effects of atmospheric aerosols on deep turbulent convective clouds using a spectral microphysics mixed-phase cumulus cloud model. Part II: Sensitivity study. *J. Atmospheric Sciences*, **61**, 2963–2982.

Khairoutdinov, M., D. Randall, and C. DeMott: 2005: Simulations of the Atmospheric General Circulation Using a Cloud-Resolving Model as a Superparameterization of Physical Processes, *J. Atmospheric Sciences*, **62**(7), 2136–2154, doi:10.1175/JAS3453.1.

Kidson, J.W., and C.S. Thompson, 1998: Comparison of statistical and model-based downscaling techniques for estimating local climate variations. *J. Climate*, **11**, 735–753.

Kiehl, J.T., 2007: Twentieth Century climate model response and climate sensitivity. *Geophysical Research Letters*, **34**, L22710, doi:10.1029/2007GL031383.

Kiehl, J.T., et al., 2006: The climate sensitivity of the Community Climate System Model: CCSM3. *J. Climate*, **19**, 2584–2596.

Kiehl, J.T., et al., 1996: *Description of the NCAR Community Climate Model (CCM3). NCAR Technical Note*. NCAR/TN-420+STR, National Center for Atmospheric Research, Boulder, Colo., 152 pp. [Available from NCAR; P.O. Box 3000; Boulder, CO 80305]

Kiktev, D., et al., 2003: Comparison of modeled and observed trends in indices of daily climate extremes. *J. Climate*, **16**, 3560–3571.

Kim, J., et al., 2005: The effects of the Gulf of California SSTs on warm-season rainfall in the southwestern United States and northwestern Mexico: A regional model study. *J. Climate*, **18**, 4970–4992.

Kim, J., and J.E. Lee, 2003: A multiyear regional climate hindcast for the western United States using the mesoscale atmospheric simulation model. *J. Hydrometeorology*, **4**(5), 878–890.

Kimoto, M., 2005: Simulated change of the East Asian circulation under global warming scenario. *Geophysical Research Letters*, **32**, L16701.

King, J.C., and J. Turner, 1997: *Antarctic Meteorology and Climatology*. Cambridge University Press, Cambridge, U.K., 256 pp.

Kirschbaum, M.U.F., 2000: Will changes in soil organic carbon act as a positive or negative feedback on global warming? *Biogeochemistry*, **48**, 21–51.

Kitoh, A., and T. Uchiyama, 2006: Changes in onset and withdrawal of the East Asian summer rainy season by multi-model global warming experiments. *J. Meteorological Society Japan*, **84**, 247–258.

Kleidon, A., 2004: Global datasets of rooting zone depth inferred from inverse methods. *J. Climate*, **17**, 2714–2722.

Klein, S.A., and C. Jakob, 1999: Validation and sensitivities of frontal clouds simulated by the ECMWF model. *Monthly Weather Review*, **127**, 2514–2531.

Klemp, J.B., and R. Wilhelmson, 1978: The simulation of three-dimensional convective storm dynamics. *J. Atmospheric Sciences*, **35**, 1070–1096.

Knowles, N., and D.R. Cayan, 2004: Elevational dependence of projected hydrologic changes in the San Francisco estuary and watershed. *Climatic Change*, **62**, 319–336.

Knutson, T.R., et al., 2007: Simulation of recent multidecadal increase of Atlantic hurricane activity using an 18-km regional model. *Bulletin American Meteorological Society*, **88**, 1549–1565.

Knutson, T.R., et al., 2006: Assessment of Twentieth Century regional surface temperature trends using the GFDL CM2 coupled models. *J. Climate*, **19**(9), 1624–1651.

Knutson, T.R., and S. Manabe, 1998: Model assessment of decadal variability and trends in the tropical Pacific Ocean. *J. Climate*, **11**, 2273–2296.

Knutti, R., et al., 2006: Constraining climate sensitivity from the seasonal cycle in surface temperature. *J. Climate*, **19**(17), 4224–4233.

Koren, V., et al., 1999: A parameterization of snowpack and frozen ground intended for NCEP weather and climate models. *J. Geophysical Research*, **104**(D16), 19569–19585.

Kraus, E.B., and J.S. Turner, 1967: A one-dimensional model of the seasonal thermocline. II: The general theory and its consequences. *Tellus*, **19**, 98–105.

Kravtsov, S.V., and C. Spannagle, 2007: Multi-decadal climate variability in observed and modeled surface temperatures. *J. Climate*, submitted.

Krinner, G., 2003: Impact of lakes and wetlands on boreal climate. *J. Geophysical Research*, **108**(D16), 4520.

Krinner, E., 1991: Northern peatlands: Role in the carbon cycle and probable responses to climatic warming. *Ecological Applications*, **1**, 182–195.

Kripalani, R.H., J.H. Oh, and H.S. Chaudhari, 2007: Response of the East Asian summer monsoon to doubled atmospheric CO_2: Coupled climate models simulations and projections under IPCC AR4. *Theoretical Applied Climatology*, **87**, 1–28.

Kripalani, R.H., et al., 2007: South Asian summer monsoon precipitation variability: Coupled climate simulations and projections under IPCC AR4. *Theoretical Applied Climatology*, doi:10.1007/s00704-006-0282-0.

Krueger, S.K., 1988: Numerical simulation of tropical cumulus clouds and their interaction with the subcloud layer. *J. Atmospheric Sciences*, **45**, 2221–2250.

Kuang, Z., and D.L. Hartmann, 2007: Testing the fixed anvil temperature (FAT) hypothesis in a cloud-resolving model. *J. Climate*, **20**, 2051–2057.

Kunkel, K.E., et al., 2006: Can CGCMs simulate the Twentieth Century "warming hole" in the central United States? *J. Climate*, **19**, 4137–4153.

Kunkel, K.E., et al., 2004: Temporal variations in frost-free season in the United States: 1895–2000. *Geophysical Research Letters*, **31**, L03201.

Kunkel, K.E., et al., 2002: Observations and regional climate model simulations of heavy precipitation events and seasonal anomalies: A comparison. *J. Hydrometeorology*, **3**, 322–334.

Kunkel, K.E., et al., 1996: The July 1995 heat wave in the Midwest: A climatic perspective and critical weather factors. *Bulletin American Meteorological Society*, **77**, 1507–1518.

Kuo, H.L., 1974: Further studies of the parameterization of the influence of cumulus convection on large-scale flow. *J. Atmospheric Sciences*, **31**, 1232–1240.

Laprise, R., 2003: Resolved scales and nonlinear interactions in limited-area models. *J. Atmospheric Sciences*, **60**, 768–779.

Large, W., J.C. McWilliams, and S.C. Doney, 1994: Oceanic vertical mixing: A review and a model with a nonlocal boundary mixing parameterization. *Reviews Geophysics*, **32**, 363–403.

Latif, M., et al., 2001: ENSIP: The El Niño simulation intercomparison project. *Climate Dynamics*, **18**, 255–272.

Lau, K.-M., et al., 2006: A multi-model study of the 20th Century simulations of Sahel drought from the 1970s to 1990s. *J. Geophysical Research*, **111**(D0711).

Lawrence, D.M., and A.G. Slater, 2005: A projection of severe near-surface permafrost degradation during the 21st Century. *Geophysical Research Letters*, **32**, L24401.

Le Treut, H., Z.X. Li, and M. Forichon, 1994: Sensitivity of the LMD general circulation model to greenhouse forcing associated with two different cloud water parameterizations. *J. Climate*, **7**, 1827–1841.

Le Treut, H., and Z.X. Li, 1991: Sensitivity of an atmospheric general circulation model to prescribed SST changes: Feedback effects associated with the simulation of cloud optical properties. *Climate Dynamics*, **5**, 175–187.

Leaman, K., R. Molinari, and P. Vertes, 1987: Structure and variability of the Florida current at 27N: April 1982–July 1984. *J. Physical Oceanography*, **17**, 565–583.

Lean, J., J. Beer, and R. Bradley, 1995: Reconstruction of solar irradiance since 1610: Implications for climate change. *Geophysical Research Letters*, **22**, 3195–3198.

Lee, H.-C., A. Rosati, and M.J. Spelman, 2006: Barotropic tidal mixing effects in a coupled climate model: Ocean conditions in the northern Atlantic. *Ocean Modelling*, **11**, 464–477.

Leith, C., 1975: Climate response and fluctuation dissipation. *J. Atmospheric Sciences*, **32**, 2022–2026.

Leung, L.R., et al., 2004: Mid-century ensemble regional climate change scenarios for the western United States. *Climate Change*, **62**, 75–113.

Leung, L.R., and Y. Qian, 2003: The sensitivity of precipitation and snowpack simulations to model resolution via nesting in regions of complex terrain. *J. Hydrometeorology*, **4**, 1025–1043.

Leung, L.R., Y. Qian, and X. Bian, 2003: Hydroclimate of the western United States based on observations and regional climate simulation of 1981–2000. Part I: Seasonal statistics. *J. Climate*, **16**(12), 1892–1911.

Leung, L.R., and M.S. Wigmosta, 1999: Potential climate change impacts on mountain watersheds in the Pacific Northwest. *J. American Water Resources Association*, **35**, 1463–1471.

Levitus, S., et al., 2001: Anthropogenic warming of Earth's climate system. *Science*, **292**, 267–270.

Li, K.Y., et al., 2006: Root-water-uptake based upon a new water stress reduction and an asymptotic root distribution function. *Earth Interactions*, **10**, Art. No. 14.

Li, W., R. Fu, and R.E. Dickinson, 2006: Rainfall and its seasonality over the Amazon in the 21st Century as assessed by the coupled models for the IPCC AR4. *J. Geophysical Research*, **11**, D02111.

Li, X., and T. Koike. 2003: Frozen soil parameterization in SiB2 and its validation with GAME-Tibet observations. *Cold Regions Science Technology*, **36**, 165–182.

Li, X., et al., 2001: A comparison of two vertical-mixing schemes in a Pacific Ocean general circulation model. *J. Climate*, **14**, 1377–1398.

Li, Z.X., 1999: Ensemble atmospheric GCM simulation climate interannual variability from 1979 to 1994. *J. Climate*, **12**, 986–1001.

Liang, X.-Z., et al., 2006: Regional climate model downscaling of the U.S. summer climate and future change. *J. Geophysical Research*, **111**, D10108.

Liang, X.-Z., et al., 2004: Regional climate model simulation of summer precipitation diurnal cycle over the United States. *Geophysical Research Letters*, **31**, L24208.

Liang, X., Z. Xie, and M. Huang, 2003: A new parameterization for surface and groundwater interactions and its impact on water budgets with the variable infiltration capacity (VIC) land surface model. *J. Geophysical Research*, **108**, 8613.

Liang, X.-Z., K.E. Kunkel, and A.N. Samel, 2001: Development of a regional climate model for U.S. midwest applications. Part 1: Sensitivity to buffer zone treatment. *J. Climate*, **14**, 4363–4378.

Libes, S.M., 1992: *An Introduction to Marine Biogeochemistry*. New York, Wiley, 734 pp.

Lin, J.L., et al., 2006: Tropical intraseasonal variability in 14 IPCC AR4 climate models. Part I: Convective signals. *J. Climate*, **19**(12), 2665–2690.

Lin, J.L., et al., 2004: Stratiform precipitation, vertical heating profiles, and the Madden–Julian Oscillation. *J. Atmospheric Sciences*, **61**, 296–309.

Lin, S.J., and R.B. Rood, 1996: Multidimensional flux-form semi-lagrangian transport schemes. *Monthly Weather Review*, **124**, 2046–2070.

Lin, W.Y., and M.H. Zhang, 2004: Evaluation of clouds and their radiative effects simulated by the NCAR Community Atmospheric Model CAM2 against satellite observations. *J. Climate*, **17**, 3302–3318.

Lindzen, R.S., M.-D. Chou, and A. Y. Hou, 2001: Does Earth have an adaptive infrared iris? *Bulletin American Meteorological Society*, **82**, 417–432.

Lindzen, R.S., and C. Giannitsis, 1998: On the climatic implications of volcanic cooling. *J. Geophysical Research*, **103**(D6), 5929–5941.

Lindzen, R.S., 1994: Climate dynamics and global change. *Annual Review Fluid Mechanics*, **26**, 353–378, doi:10.1146/annurev.fl.26.010194.002033.

Lindzen, R.S., and M.S. Fox-Rabinovitz, 1989: Consistent vertical and horizontal resolution. *Monthly Weather Review*, **117**, 2575–2583.

Liston, G.E., 2004: Representing subgrid snow cover heterogeneities in regional and global models. *J. Climate*, **17**, 1381–1397.

Liu, Z., 1998: On the role of ocean in the transient response of tropical climatology to global warming. *J. Climate*, **11**, 864–875.

Lock, A.P., et al., 2000: A new boundary layer mixing scheme. Part I: Scheme description and single-column model tests. *Monthly Weather Review*, **128**, 3187–3199.

Lock, A., 1998: The parameterization of entrainment in cloudy boundary layers. *Quarterly J. Royal Meteorological Society*, **124**, 2729–2753.

Lockwood, M., and C. Fröhlich, 2007: Recent oppositely directed trends in solar climate forcings and the global mean surface air temperature. *Proceedings Royal Society A*, **463**, 2447–2460, doi:10.1098/rspa.2007.1880.

Lofgren, B.M., 2004: A model for simulation of the climate and hydrology of the Great Lakes basin. *J. Geophysical Research*, **109**(D18), Art. No. D18108.

Lohmann, U., and E. Roeckner, 1996: Design and performance of a new cloud microphysics scheme developed for the ECHAM4 general circulation model. *Climate Dynamics*, **12**, 557–572.

Lorenz, P., and D. Jacob, 2005: Influence of regional scale information on the global circulation: A two-way nesting climate simulation. *Geophysical Research Letters*, **32**, Art. No. L18706.

Lucarini, V., et al., 2006: Intercomparison of the Northern Hemisphere winter mid-latitude atmospheric variability of the IPCC models. *Climate Dynamics*, **28**(7–8), 829–848.

Lumpkin, R., and K. Speer, 2003: Large-scale vertical and horizontal circulation in the North Atlantic Ocean. *J. Physical Oceanography*, **33**, 1902–1920.

Luo, L.F., et al., 2003: Effects of frozen soil on soil temperature, spring infiltration, and runoff: Results from the PILPS 2(d) experiment at Valdai, Russia. *J. Hydrometeorology*, **4**, 334–351.

Lynch, A.H., J.A. Maslanik, and W.L. Wu, 2001: Mechanisms in the development of anomalous sea ice extent in the western Arctic: A case study. *J. Geophysical Research*, **106**(D22), 28097–28105.

Lynch, A.H., et al., 1995: Development of a regional climate model of the western Arctic. *J. Climate*, **8**, 1555–1570.

Lynn, B.H., et al., 2005: Spectral (bin) microphysics coupled with a mesoscale model (MM5). Part II: Simulation of a CaPE rain event with a squall line. *Monthly Weather Review*, **133**, 59–71.

Maak, K., and H. von Storch, 1997: Statistical downscaling of monthly mean temperature to the beginning of flowering of *Galanthus nivalis* L. in Northern Germany. *International J. Biometeorology*, **41**(1), 5–12.

Malevsky–Malevich, S.P., et al., 1999: The evaluation of climate change influence on the permafrost season soil thawing regime. *Contemporary Investigation Main Geophysical Observatory*, **1**, 33–50 (in Russian).

Maltrud, M.E., et al., 1998: Global eddy-resolving ocean simulations driven by 1985–1995 atmospheric winds. *J. Geophysical Research*, **103**(C13), 30, 825–30, 853.

Manabe, S., et al., 1991: Transient responses of a coupled ocean-atmosphere model to gradual changes of atmospheric CO_2. Part I: Annual mean response. *J. Climate*, **4**, 785–818.

Manabe, S., and A. Broccoli, 1985: A comparison of climate model sensitivity with data from the last glacial maximum. *J. Atmospheric Sciences*, **42**, 2643–2651.

Manabe, S., and R.T. Wetherald, 1975: The effects of doubling the CO_2 concentration on the climate of a general circulation model. *J. Atmospheric Sciences*, **32**, 3–15.

Manabe, S., 1969: Climate and the ocean circulation. 1: The atmospheric circulation and hydrology of the Earth's surface. *Monthly Weather Review*, **97**, 739–774.

Manabe, S., and R.T. Wetherald, 1967: Thermal equilibrium of the atmosphere with a given distribution of relative humidity. *J. Atmospheric Sciences*, **24**(3), 241–259.

Manabe, S., J. Smagorinsky, and R.F. Strickler, 1965: Simulated climatology of a general circulation model with a hydrological cycle. *Monthly Weather Review*, **93**, 769–798.

Martin, G.M., et al., 2000: A new boundary layer mixing scheme. Part II: Tests in climate and mesoscale models. *Monthly Weather Review*, **128**, 3200–3217.

Martin, J.H., et al., 1994: Testing the iron hypothesis in ecosystems of the equatorial Pacific Ocean. *Nature*, **371**, 123–129.

Martin, J.H., 1991: Iron as a limiting factor in oceanic productivity. In *Primary Productivity and Biogeochemical Cycles in the Sea*, ed. P.G. Falkowski and A.D. Woodhead, 123–137. Plenum Press, New York.

Maurer, E. P., et al., 2002: A long-term hydrologically-based data set of land surface fluxes and states for the conterminous United States. *J. Climate*, **15**, 3237–3251.

Maxwell, R.M., and N.L. Miller, 2005: Development of a coupled land surface and groundwater model. *J. Hydrometeorology*, **6**, 233–247.

May, W., and E. Roeckner, 2001: A time-slice experiment with the ECHAM4 AGCM at high resolution: The impact of horizontal resolution on annual mean climate change. *Climate Dynamics* **17**(5–6), 407–420.

McCreary, J., and P. Lu, 1994: Interaction between the subtropical and equatorial ocean circulation—The subtropical cell. *J. Physical Oceanography*, **24**, 466–497.

McCulloch, M.T., and T. Esat, 2000: The coral record of the last interglacial sea levels and sea surface temperatures. *Chemical Geology*, **169**, 107–129.

McFarlane, N.A., 1987: The effect of orographically excited gravity wave drag on the general circulation of the lower stratosphere and troposphere. *J. Atmospheric Sciences*, **44**, 1775–1800.

McGregor, J.L., 1999: Regional Modelling at CAR: Recent Developments, 43–48 in *Parallel Computing in Meteorology and Oceanography*. BMRC Research Report No. 75, Bureau of Meteorology, Melbourne, Australia.

McGregor, J.L., 1997: Regional climate modelling. *Meteorology Atmospheric Physics*, **63**, 105–117.

McPhaden, M.J., et al., 1998: The Tropical Ocean Global Atmosphere (TOGA) observing system: A decade of progress. *J. Geophysical Research*, **103**, 14169–14240.

Mearns, L.O., 2003: Issues in the impacts of climate variability and change on agriculture—Applications to the southeastern United States. *Climate Change*, **60**, 1–6.

Mearns, L.O., et al., 2003: Climate scenarios for the southeastern U.S. based on GCM and regional model simulations. *Climate Change*, **60**, 7–35.

Mearns, L.O., et al., 1999: Comparison of climate change scenarios generated from regional climate model experiments and statistical downscaling. *J. Geophysical Research*, **104**, 6603–6621.

Mechoso, C.R., et al., 1995: The seasonal cycle over the tropical Pacific in coupled ocean-atmosphere general circulation models. *Monthly Weather Review*, **123**, 2825–2838.

Meehl, G.A., et al., 2006: Climate change projections for the Twenty-First Century and climate change commitment in the CCSM3. *J. Climate*, **19**, 2597–2616.

Meehl, G.A., et al., 2005: How much more will global warming and sea level rise? *Science*, **307**, 1769–1772.

Meehl, G.A., and C. Tebaldi, 2004: More intense, more frequent, and longer lasting heat waves in the 21st Century. *Science*, **305**, 994–997.

Meehl, G.A., C. Tebaldi, and D. Nychka, 2004: Changes in frost days in simulations of Twenty-First Century climate. *Climate Dynamics*, **23**, 495–511.

Mellor, G.L., and T. Yamada, 1982: Development of a turbulence closure model for geophysical fluid problems. *Reviews Geophysics Space Physics*, **20**, 851–875.

Mellor, G.L., and T. Yamada, 1974: A hierarchy of turbulent closure models for planetary boundary layers. *J. Atmospheric Sciences*, **31**, 1791–1806.

Menon, S., and A. Del Genio, 2007: Evaluating the Impacts of Carbonaceous Aerosols on Clouds and Climate, 34–48 in *Human-Induced Climate Change: An Interdisciplinary Assessment*, ed. M.E. Schlesinger et al. Cambridge University Press, Cambridge, U.K., and New York, N.Y.

Merryfield, W.J., 2006: Changes to ENSO under CO_2 doubling in a multi-model ensemble. *J. Climate*, **19**, 4009–4027.

Miguez-Macho, G., G.L. Stenchikov, and A. Robock, 2005: Regional climate simulations over North America: Interaction of local processes with improved large-scale flow. *J. Climate*, **18**, 1227–1246.

Miller, D.A., and R.A. White, 1998: A coterminous United States multilayer soil characteristics dataset for regional climate and hydrology modeling. *Earth Interactions*, **2**, 1–26.

Miller, R.L., G.A. Schmidt, and D.T. Shindell, 2006: Forced variations of annular modes in the 20th Century IPCC AR4 simulations. *J. Geophysical Research*, **111**, D18101, doi:10.1029/2005JD006323.

Min, S.-K., and A. Hense, 2007: A Bayesian assessment of climate change using multi-model ensembles. Part II: Regional and seasonal mean surface temperature. *J. Climate*, **20**(12), 1769–1790.

Min, S.-K., and A. Hense, 2006: A Bayesian assessment of climate change using multi-model ensembles. Part I: Global mean surface temperature. *J. Climate*, **19**, 3237– 3256.

Minschwaner, K., E.D. Essler, and P. Sawaengphokhai, 2006: Multi-model analysis of the water vapor feedback in the tropical upper troposphere. *J. Climate*, **19**, 5455–5464.

Minschwaner, K., and A.E. Dessler, 2004: Water vapor feedback in the tropical upper troposphere: Model results and observations. *J. Climate*, **17**(6), 272–1282.

Mirocha, J.D., B. Kosovic, and J.A. Curry, 2005: Vertical heat transfer in the lower atmosphere over the Arctic Ocean during clear-sky periods. *Boundary-Layer Meteorology*, **117**, 37–71.

Mitchell, J.F.B., et al., 1999: Towards the construction of climate change scenarios. *Climatic Change*, **41**(3–4), 547–581.

Mitchell, J.F.B., C.A. Senior, and W.J. Ingram, 1989: CO_2 and climate: A missing feedback? *Nature*, **341**, 132–134.

Mitchell, T., 2003: Pattern scaling: An examination of the accuracy of the technique for describing future climates. *Climatic Change*, **60**, 217–242.

Miura, H., et al., 2005: A climate sensitivity test using a global cloud resolving model under an aqua planet condition. *Geophysical Research Letters*, **32**, L19717.

Mo, K.C., et al., 2005: Impact of model resolution on the prediction of summer precipitation over the United States and Mexico. *J. Climate*, **18**, 3910–3927.

Moorthi, S., and M.J. Suarez, 1992: Relaxed Arakawa-Schubert: A parameterization of moist convection for general circulation models. *Monthly Weather Review*, **120**, 978–1002.

Morel, A., and D. Antoine, 1994: Heating rate within the upper ocean in relation to its bio-optical state. *J. Physical Oceanography*, **24**, 1652–1665.

Morrison, H., and J.O. Pinto, 2005: Mesoscale modeling of springtime Arctic mixed-phase stratiform clouds using two-moment bulk microphysics scheme. *J. Atmospheric Sciences*, **62**, 3683–3704.

Murphy, J.M., et al., 2004: Quantification of modelling uncertainties in a large ensemble of climate change simulations. *Nature*, **430**(7001), 768–772.

Murray, R.J., 1996: Explicit generation of orthogonal grids for ocean models. *J. Computational Physics*, **126**, 251–273.

Nadelhoffer, K.J., et al., 1999: Nitrogen makes a minor contribution to carbon sequestration in temperate forests. *Nature*, **398**, 145–148.

Najjar, R.G., et al., 2007: Impact of circulation on export production, dissolved organic matter, and dissolved oxygen in the ocean: Results from OCMIP-2. *Global Biogeochemistry Cycles*, **21**, GB3007, doi:10.1029/2006GB002857.

NARCCAP (North American Regional Climate Change Assessment), cited as 2007: www.narccap.ucar.edu/.

Neelin, J.D., et al., 1992: Tropical air-sea interaction in general circulation models. *Climate Dynamics*, **7**, 73–104.

Niu, G.Y., and Z.L. Yang, 2006: Effects of frozen soil on snowmelt runoff and soil water storage at a continental scale. *J. Hydrometeorology*, **7**, 937–952.

Nordeng, T.E., 1994: Extended Versions of the Convective Parameterization Scheme at ECMWF and Their Impact on the Mean and Transient Activity of the Model in the Tropics. *Technical Memorandum 206, European Center for Medium Range Weather Forecasting (ECMWF)*, Reading, U.K.

Norris, J., and C.P. Weaver, 2001: Improved techniques for evaluating GCM cloudiness applied to the NCAR CCM3. *J. Climate*, **14**, 2540–2550.

Nowak, R.S., D.S. Ellsworth, and S.D. Smith, 2004: Tansley review: Functional responses of plants to elevated atmospheric CO_2—Do photosynthetic and productivity data from FACE experiments support early predictions? *New Phytologist*, **162**, 253–280.

Oglesby, R.J., and B. Saltzman, 1992: Equilibrium climate statistics of a general circulation model as a function of atmospheric carbon dioxide. Part I: Geographic distributions of primary variables. *J. Climate*, **5**(1), 66–92.

Ohlmann, J.C., 2003: Ocean radiant heating in climate models. *J. Climate*, **16**, 1337–1351.

Olesen, J.E., et al., 2007: Uncertainties in projected impacts of climate change on European agriculture and terrestrial ecosystems based on scenarios from regional climate models. *Climatic Change*, **81**, 123–143.

Oleson, K.W., et al., 2004: Technical Description of the Community Land Model (CLM). *NCAR Technical Note*. NCAR/TN-461+STR, National Center for Atmospheric Research, Boulder, Colo., 173 pp. [Available from NCAR; P.O. Box 3000; Boulder, CO 80305]

Oouchi, K., et al., 2006: Tropical cyclone climatology in a global-warming climate as simulated in a 20- km mesh global atmospheric model: Frequency and wind intensity analyses. *J. Meteorological Society Japan*, **84**, 259–276.

Oren, R., et al., 2001: Soil fertility limits carbon sequestration by forest ecosystems in a CO_2–enriched atmosphere. *Nature*, **411**, 469–477.

Ott, L.E., et al., 2007: The effects of lightning NOx production during the 21 July EULINOX storm studied with a 3-D cloud-scale chemical transport mode. *J. Geophysical Research*, **112**, D05307, doi:10.1029/2006JD007365.

Our Changing Planet: The FY 1992 U.S. Global Change Research Program. 1991. A Report by the Committee on Earth and Environmental Sciences. A Supplement to the U.S. President's Fiscal Year 1992 Budget.

Overgaard, J., D. Rosbjerg, and M.B. Butts, 2006: Land-surface modelling in hydrological perspective–A review. *Biogeoscience*, **3**, 229–241.

Pacanowski, R.C., and S.G.H. Philander, 1981: Parameterization of vertical mixing in numerical models of tropical oceans. *J. Physical Oceanography*, **11**, 1443–1451.

Paegle, J., K.C. Mo, and J. Nogués-Paegle, 1996: Dependence of simulated precipitation on surface evaporation during the 1993 United States summer floods. *Monthly Weather Review*, **124**, 345–361.

Pan, Z., et al., 2004: Altered hydrologic feedback in a warming climate introduces a "warming hole." *Geophysical Research Letters*, **31**, L17109, doi:10.1029/2004GL020528.

Pan, Z., et al., 2001: Evaluation of uncertainties in regional climate change simulations. *J. Geophysical Research*, **106**, 17735–17752.

Parkinson, C.L., K.Y. Vinnikov, and D.J. Cavalieri, 2006a: Evaluation of the simulation of the annual cycle of Arctic and Antarctic. *J. Geophysical Research*, **111**, C07012.

Parkinson, C.L., K.Y. Vinnikov, and D.J. Cavalieri, 2006b: Correction to evaluation of the simulation of the annual cycle of Arctic and Antarctic sea ice coverages by 11 major global climate models. *J. Geophysical Research*, **111**, C12009, doi:10.1029/2006JC003949.

Paulson, C.A., and J.J. Simpson, 1977: Irradiance measurements in the upper ocean. *J. Applied Oceanography*, **7**, 952–956.

Pavolonis, M., J.R. Key, and J.J. Cassano, 2004: Study of the Antarctic surface energy budget using a polar regional atmospheric model forced with satellite-derived cloud properties. *Monthly Weather Review*, **132**, 654–661.

Pawson, S., et al., 2000: The GCM-Reality Intercomparison Project for SPARC GRIPS): Scientific issues and initial results. *Bulletin American Meteorological Society*, **81**, 781–796.

Payne, A.J., et al., 2004: Recent dramatic thinning of largest West Antarctic ice stream triggered by oceans. *Geophysical Research Letters*, **31**, L23401.

Payne, J.T., et al., 2004: Mitigating the effects of climate change on the water resources of the Columbia River basin. *Climatic Change*, **62**, 233–256.

Peltier, W.R., 2004: Global glacial isostasy and the surface of the ice-age earth: The ice-5G (VM2) model and GRACE. *Annual Review Earth Planetary Science*, **32**, 111–149.

Philander, S.G.H., 1990: *El Niño, La Niña, and the Southern Oscillation*. Academic Press, San Diego, Calif., 293 pp.

Philander, S.G.H., and R.C. Pacanowski, 1981: The oceanic response to cross-equatorial winds (with application to coastal upwelling in low latitudes). *Tellus*, **33**, 201–210.

Philip, S.Y., and G. J. van Oldenborgh. 2006: Shifts in ENSO coupling processes under global warming. *Geophysical Research Letters*, **33**, L11704.

Phillips, T.J., et al., 2004: Evaluating parameterizations in general circulation models: Climate simulation meets weather prediction. *Bulletin American Meteorological Society*, doi:10.1175/BAMS-85-12-1903.

Piani, C., et al., 2005: Constraints on climate change from a multi-thousand member ensemble of simulations. *Geophysical Research Letters*, **32**, L23825.

Pierce, D.W., et al., 2006: Three-dimensional tropospheric water vapor in coupled climate models compared with observations from the AIRS satellite system. *Geophysical Research Letters*, **33**, L21701.

Pierce, D.W., 2004: Future change in biological activity in the north Pacific due to anthropogenic forcing of the physical environment. *Climatic Change*, **62**, 389–418.

Pincus, R., H.W. Barker, and J.J. Morcrette, 2003: A fast, flexible, approximate technique for computing radiative transfer in inhomogeneous cloud fields. *J. Geophysical Research*, doi:10.1209/2002JD0033222003.

Pinto, J.O., J.A. Curry, and J.M. Intrieri, 2001: Cloud-aerosol interactions during autumn over Beaufort Sea. *J. Geophysical Research*, **106**, 15077–15097.

Pitman, A.J., et al., 1999: Uncertainty in the simulation of runoff due to the parameterization of frozen soil moisture using the GSWP methodology. *J. Geophysical Research*, **104**, 16879–16888.

Pitsch, H., 2006: Large-eddy simulation of turbulent combustion. Annual *Review Fluid Dynamics*, **38**, 453–482.

Plummer, D.A., et al., 2006: Climate and climate change over North America as simulated by the Canadian Regional Climate Model. *J. Climate*, **19**, 3112–3132.

Polvani, L.M., R.K. Scott, and S.J. Thomas: 2004: Numerically converged solutions of the global primitive equations for testing the dynamical core of atmospheric GCMs. *Monthly Weather Review*, **132**, 2539–2552.

Pope, V.D., et al., 2000: The impact of new physical parameterizations in the Hadley Centre Climate Model—HadAM3. *Climate Dynamics*, **16**, 123–146.

Prentice, I.C., et al., 2001: The Carbon Cycle and Atmospheric Carbon Dioxide. Chapter 3 in *Climate Change 2001: The Scientific Basis: Contribution of Working Group 1 to the Third Assessment Report of the Intergovernmental Panel on Climate Change*, ed. J.T. Houghton et al. Cambridge University Press, Cambridge, U.K., and New York, N.Y.

Privé, N.C., and R.A. Plumb, 2007: Monsoon dynamics with interactive forcing. Part I: Axisymmetric studies. *J. Atmospheric Sciences,* **64**(5), 1417–1430.

Qian, J.-H., W.-K. Tao, and K.-M. Lau, 2004: Mechanisms for torrential rain associated with the mei-yu development during SCSMEX 1998. *Monthly Weather Review*, **132**, 3–27.

Qian, J.-H., F. Giorgi, and M.S. Fox–Rabinovitz, 1999: Regional stretched grid generation and its application to the NCAR RegCM. *J. Geophysical Research*, **104**(D6), 6501–6514.

Qu, X., and A. Hall, 2006: Assessing snow albedo feedback in simulated climate change. *J. Climate*, **19**(11), 2617–2630.

Raisanen, J., 2002: CO_2–induced changes in interannual temperature and precipitation variability in 19 CMIP experiments. *J. Climate*, **15**, 2395–2411.

Raisanen, J., and T.N. Palmer, 2001: A probability and decision-model analysis of a multimodel ensemble of climate change simulations. *J. Climate*, **14**, 3212–3226.

Rajagopalan, B., U. Lall, and M.A. Cane, 1997: Anomalous ENSO occurrences: An alternate view. *J. Climate*, **10**(9), 2351–2357.

Ramankutty, N., et al., 2002: The global distribution of cultivable lands: Current patterns and sensitivity to possible climate change. *Global Ecology Biogeography*, **11**, 377–392.

Ramaswamy, V., et al., 2001: Radiative Forcing Climate Change, 349–416, in *Climate Change 2001: The Scientific Basis*, ed. J.T. Houghton, et al. Cambridge University Press, Cambridge, U.K.

Randall, D.A., et al., 2007: Climate Models and Their Evaluation, Chapter 8, in *Climate Change 2007. The Fourth Scientific Assessment*, ed. S. Solomon et al. Intergovernmental Panel on Climate Change (IPCC) (www.ipcc.ch).

Randall, D.A., et al., 2003: Breaking the cloud parameterization deadlock. *Bulletin American Meteorological Society*, **84**, 1547–1564, doi:10.1175/BAMS-84-11-1547.

Randall, D.A., et al., 2000: Cloud Feedbacks in *Frontiers in Science: Climate Modeling*, ed. J.T. Kiehl and V. Ramanathan. Proceedings of a symposium in honor of Robert D. Cess.

Raper, S.C.B., J.M. Gregory, and R.J. Stouffer, 2002: The role of climate sensitivity and ocean heat uptake on AOGCM transient temperature response. *J. Climate*, **15**, 124–130.

Raphael, M.N., and M.M. Holland, 2006: Twentieth Century simulation of the Southern Hemisphere climate in coupled models. Part 1: Large-scale circulation variability. *Climate Dynamics*, **26**, 217–228.

Rasch, P.J., and J.E. Kristjánsson, 1998: A comparison of the CCM3 model climate using diagnosed and predicted condensate parameterizations. *J. Climate*, **11**, 1587–1614.

Rasmussen, E.M., and J.M. Wallace, 1983: Meteorological aspects of the El Niño/Southern Oscillation. *Science*, **222**, 1195–2002.

Rawlins, M.A., et al., 2003: Simulating pan-Arctic runoff with a macro-scale terrestrial water balance model. *Hydrology Proceedings*, **17**, 2521–2539.

Rayner, N.A., et al., 2003: Global analyses of sea surface temperature, sea ice, and night marine air temperature since the late Nineteenth Century. *J. Geophysical Research*, **108**(D14), 4407.

Reichler, T., and J. Kim, 2008: How well do coupled models simulate today's climate? *Bulletin American Meteorological Society*, **89**(3), doi:10.1175/BAMS-89-3-303, (www.met.utah.edu/reichler/publications/papers/Reichler_07_BAMS_CMIP.pdf).

Rignot, E., and P. Kanagaratnam, 2006: Changes in the velocity structure of the Greenland ice sheet. *Science*, **311**, 986–990.

Ringer, M.A., and R.P. Allan, 2004: Evaluating climate model simulations of tropical clouds. *Tellus*, **56A**, 308–327.

Rinke, A., et al., 2006: Evaluation of an ensemble of Arctic regional climate models: Spatiotemporal fields during the SHEBA year. *Climate Dynamics*, doi:10.1007/s00382-005-0095-3.

Roads, J., S.-C. Chen, and M. Kanamitsu, 2003: U.S. regional climate simulations and seasonal forecasts. *J. Geophysical Research*, **108**(D16), Art. No. 8606.

Roads, J.O., et al., 1999: Surface water characteristics in the NCEP Global Spectral Model and reanalysis. *J. Geophysical Research*, **4**(D16), 19307–19327.

Roberts, M.J., and R. Wood, 1997: Topographic sensitivity studies with a Bryan-Cox-type ocean model. *J. Physical Oceanography*, **27**, 823–836.

Roe, G.H., and M.B. Baker, 2007: Why is climate sensitivity so unpredictable? *Science*, **318**(5850), 629–632, doi:10.1126/science.1144735.

Roeckner, E., et al., 2006: Sensitivity of simulated climate to horizontal and vertical resolution in the ECHAM5 atmosphere model. *J. Climate*, **19**, 3771–3791.

Roeckner, E., et al., 1996: The atmospheric general circulation model ECHAM-4: Model description and simulation of present-day climate. Report 128, Max-Planck-Institut für Meteorologie, Hamburg, Germany.

Roeckner, E., et al., 1987: Cloud optical depth feedbacks and climate modelling. *Nature*, **329**, 138–140.

Root, T.L., and S.H. Schneider, 1993: Can large-scale climatic models be linked with multiscale ecological studies? *Conservation Biology*, **7**(2), 256–270.

Ropelewski, C.F., and M.S. Halpert, 1987: Global and regional scale precipitation patterns associated with the El Niño Southern Oscillation. *Monthly Weather Review*, **115**, 1606–1626.

Rosenzweig, C., et al., 2002: Increased crop damage in the U.S. from excess precipitation under climate change. *Global Environmental Change*, **12**, 197–202.

Rothstein, L.M., et al., 2006: Modeling ocean ecosystems: The PARADIGM program. *Oceanography*, **19**, 22–51.

Rotstayn, L.D., and U. Lohmann, 2002: Tropical rainfall trends and the indirect aerosol effect. *J. Climate*, **15**, 2103–2116.

Rotstayn, L.D., 1997: A physically based scheme for the treatment of stratiform clouds and precipitation in large-scale models. Part I: Description and evaluation of microphysical processes. *Quarterly J. Royal Meteorological Society*, **123**, 1227–1282.

Rowell, D.P., et al., 1992: Modelling the influence of global sea surface temperatures on the variability and predictability of seasonal Sahel rainfall. *Geophysical Research Letters*, **19**, 905–908.

Ruiz-Barradas, A., and S. Nigam, 2006: IPCC's Twentieth Century climate simulations: Varied representations of North American hydroclimate variability. *J. Climate*, **19**, 4041–4058.

Rummukainen, M., et al., 2004: The Swedish Regional Climate Modelling Programme, SWECLIM: A review. *Ambio*, **33**, 176–182.

Ruosteenoja, K., H. Tuomenvirta, and K. Jylha, 2007: GCM-based regional temperature and precipitation change estimates for Europe under four SRES scenarios applying a super-ensemble pattern-scaling method. *Climatic Change*, **81**, Supplement 1, doi:10.1007/s10584-006-9222-3.

Russell, G.L., et al., 2000: Comparison of model and observed regional temperature changes during the past 40 years. *J. Geophysical Research*, **105**, 14891–14898.

Russell, G.L., J.R. Miller, and D. Rind, 1995. A coupled atmosphere-ocean model for transient climate change studies. *Atmosphere-Ocean*, **33**(4), 683–730.

Russell, J.L., R. Stouffer, and K.W. Dixon, 2007: Corrigendum. *J. Climate*, **20**, 4287.

Russell, J.L., R.J. Stouffer, and K.W. Dixon, 2006: Intercomparison of the Southern Ocean circulations in IPCC coupled model control simulations. *J. Climate*, **19**, 4560–4575.

Ryan, B.F., et al., 2000: Simulations of a cold front by cloud-resolving, limited-area, and large-scale models, and a model evaluation using in situ and satellite observations. *Monthly Weather Review*, **128**, 3218–3235.

Saji, N.H., S.-P. Xie, and T. Yamagata, 2005: Tropical Indian Ocean variability in the IPCC Twentieth Century climate simulations. *J. Climate*, **19**(17), 4397.

Saji, N.H., et al., 1999: A dipole mode in the tropical Indian Ocean. *Nature*, **401**, 360–363.

Santer, B.D., et al., 2007: Identification of human-induced changes in atmospheric moisture content. *Proceedings National Academy Sciences*, **104**, 15248–15253.

Santer, B.D., et al., 2005: Amplification of surface temperature trends and variability in the tropical atmosphere. *Science*, **309**, 1551–1556.

Santer, B.D., et al., 2001: Accounting for the effects of volcanoes and ENSO in comparisons of modeled and observed temperature trends. *J. Geophysical Research*, **106**, 28033–28059.

Santer, B.D., et al., 1990: *Developing Climate Scenarios from Equilibrium GCM Results*, Report No. 47, Max Planck Institute for Meteorology, Hamburg.

Saraf, A.K., et al., 2005: Digital Elevation Model (DEM) generation from NOAA-AVHRR night-time data and its comparison with USGS-DEM. *International J. Remote Sensing*, **26**, 3879–3887.

Sardeshmukh, P.D., and B.J. Hoskins, 1988: The generation of global rotational flow by steady idealized tropical divergence. *J. Atmospheric Sciences*, **45**, 1228–1251.

Sato, M., et al., 1993: Stratospheric aerosol optical depth, 1850–1990. *J. Geophysical Research*, **98**, 22987–22994.

Sausen, R., S. Schubert, and L. Dumenil, 1994: A model of the river runoff for use in coupled atmosphere-ocean models. *J. Hydrology*, **155**, 337–352.

Schimel, D.S., 1998: The carbon equation. *Nature*, **393**, 208–209.

Schmidt, G.A., et al., 2006: Present day atmospheric simulations using GISS ModelE: Comparison to in-situ, satellite, and reanalysis data. *J. Climate*, **19**, 153–192, doi:10.1175/JCLI3612.1.

Schmittner, A., M. Latif, and B. Schneider, 2005: Model projections of the North Atlantic thermohaline circulation for the 21st Century assessed by observations. *Geophysical Research Letters*, **32**, doi:10.1029/2005GL024368.

Schneider, S.H., and S.L. Thompson, 1981: Atmospheric CO_2 and climate: Importance of the transient response. *J. Geophysical Research*, **86**, 3135–3147.

Schneider, S.H., and C. Mass, 1975: Volcanic dust, sunspots, and temperature trends. *Science*, **190**, 741–746.

Schopf, P., et al., 2003: *Coupling Process and Model Studies of Ocean Mixing to Improve Climate Models—A Pilot Climate Process Modeling and Science Team* (a U.S. CLIVAR paper, www.usclivar.org/CPT/Ocean_mixing_whitepaper.pdf).

Schramm, J.L., et al., 1997: Modeling the thermodynamics of a sea ice thickness distribution. Part 1: Sensitivity to ice thickness resolution. *J. Geophysical Research*, **102**, 23079–23091.

Schwartz, S.E., 2007: Heat capacity, time constant, and sensitivity of Earth's climate system. *J. Geophysical Research*, **112**, D24S05, doi:10.1029/2007JD008746, 2007

Schweitzer, L., 2006: Environmental justice and hazmat transport: A spatial analysis in southern California. *Transportation Research. Part D–Transportation Environment*, **11**, 408–421.

Segal, M., et al., 1997: Small lake daytime breezes: Some observational and conceptual evaluations. *Bulletin American Meteorological Society*, **78**, 1135–1147.

Segal, M., and R.W. Arritt, 1992: Nonclassical mesoscale circulations caused by surface sensible heat-flux gradients. *Bulletin American Meteorological Society*, **73**, 1593–1604.

Sellers, P.J., et al., 1996: A revised land surface parameterization (SiB2) for atmospheric GCMs. Part 1: Model formulation. *J. Climate*, **9**, 676–705.

Sellers, P.J., et al., 1986: A simple biosphere model (SiB) for use within general-circulation models. *J. Atmospheric Sciences*, **43**, 503–531.

Semtner, A.J., 1976: A model for the thermodynamic growth of sea ice in numerical investigations of climate. *J. Physical Oceanography*, **6**, 27–37.

Seneviratne, S.I., et al., 2006: Land-atmosphere coupling and climate change in Europe. *Nature*, **443**, 205–209.

Senior, C.A., and J.F.B. Mitchell, 1996: Cloud Feedbacks in the Unified UKMO GCM, in *Climate Sensitivity to Radiative Perturbations, Physical Mechanism and Their Validation*, ed. H. Le Treut. Springer, 331 pp.

Senior, C.A., and J.F.B. Mitchell, 1993: Carbon dioxide and climate: The impact of cloud parameterization. *J. Climate*, **6**, 5–21.

Seth, A., and F. Giorgi, 1998: The effects of domain choice on summer precipitation simulation and sensitivity in a regional climate model. *J. Climate*, **11**, 2698–2712.

117

Shaw, M.R., et al., 2002: Grassland responses to global environmental changes suppressed by elevated CO_2. *Science*, **298**, 1987–1990.

Shepherd, A., and D. Wingham, 2007: Recent sea-level contributions of the Antarctica and Greenland ice sheets. *Science*, **315**, 1529–1532.

Shindell, D.T., et al., 2006: Solar and anthropogenic forcing of tropical hydrology. *Geophysical Research Letters*, **33**, L24706, doi:10.1029/2006GL027468.

Shindell, D.T., et al., 1999: Simulation of recent northern winter climate trends by greenhouse-gas forcing. *Nature*, **399**, 452–455.

Shukla, J., et al., 2006: Climate model fidelity and projections of climate change. *Geophysical Research Letters*, **33**, L07702, doi:10.1029/2005GL025579.

Siddall, M., et al., 2003: Sea-level fluctuations during the last glacial cycle. *Nature*, **423**, 853–858.

Skamarock, W.C., et al., 2005: A Description of the Advanced Research WRF Version 2. *NCAR Technical Note*. NCAR/TN–468+STR, National Center for Atmospheric Research, Boulder, Colo., 88 pp. [Available from NCAR; P.O. Box 3000; Boulder, CO 80305]

Slater, A.G., et al., 2001: The representation of snow in land surface schemes: Results from PILPS 2(d). *J. Hydrometeorology*, **2**, 7–25.

Small, C., V. Gornitz, and J.E. Cohen, 2000. Coastal hazards and the global distribution of population. *Environmental Geoscience*, **7**, 3–12.

Smethie, W.M., Jr., and R.A. Fine, 2001: Rates of North Atlantic deep water formation calculated from chlorofluorocarbon inventories. *Deep Sea Research, Part 1: Oceanographic Research Papers*, **48**, 189–215.

SMIC, 1971: *Inadvertent Climate Modification: Report of the Study of Man's Impact on Climate*, Massachusetts Institute of Technology Press, Cambridge, Mass., 308 pp.

Smith, R.D., and P.R. Gent, 2004: *Reference Manual for the Parallel Ocean Program (POP), Ocean Component of the Community Climate System Model (CCSM2.0 and 3.0)*. Technical Report LA-UR-02-2484, Los Alamos National Laboratory, Los Alamos, New Mexico (www.ccsm.ucar.edu/models/ccsm3.0/pop).

Smith, R.N.B., 1990: A scheme for predicting layer clouds and their water content in a general circulation model. *Quarterly J. Royal Meteorological Society*, **116**, 435–460.

Smith, S.J., et al., 2005: Climate change impacts for the conterminous USA, Part 1: Scenarios and context. *Climatic Change*, **69**, 7–25.

Soden, B.J., and I.M. Held, 2006: An assessment: Climate feedbacks in coupled ocean–atmosphere models. *J. Climate*, **19**, 3354–3360.

Soden, B.J., et al., 2005: The radiative signature of upper tropospheric moistening. *Science*, **310**(5749), 841–844, doi:10.1126/science.1115602.

Soden, B.J., A.J. Broccoli, and R.S. Hemler, 2004: On the use of cloud forcing to estimate cloud feedback. *J. Climate*, **17**, 3661–3665.

Soden, B.J., et al., 2002: Global cooling after the eruption of Mount Pinatubo: A test of climate feedback by water vapor. *Science*, **296**, 727–730.

Soden, B.J., 2000: The diurnal cycle of convection, clouds, and water vapor in the tropical upper troposphere. *Geophysical Research Letters*, **27**, 2173–2176.

Soden, B.J., and I.M. Held, 2000: An assessment of climate feedbacks in coupled ocean–atmosphere models. *J. Climate*, **19**, 3354–3360.

Sokolov, A.P., and P.H. Stone, 1998: A flexible climate model for use in integrated assessments. *Climate Dynamics*, **14**, 291–303.

Solman, S.A., M.N. Nunez, and P.R. Rowntree, 2003: On the evaluation of the representation of mid-latitude transients in the Southern Hemisphere by HadAM2B GCM and the impact of horizontal resolution. *Atmosfera*, **16**, 245–272.

Somerville, R.C.J., and L.A. Remer, 1984: Cloud optical thickness feedbacks in the CO_2 climate problem. *J. Geophysical Research*, **89**, 9668–9672.

Spencer, R.W., et al., 2007: Cloud and radiation budget changes associated with tropical intraseasonal oscillations. *Geophysical Research Letters*, **34**, L15707, doi:10.1029/2007GL029698.

Stainforth, D.A., et al., 2005: Uncertainty in predictions of the climate response to rising levels of greenhouse gases. *Nature*, **443**, 403–406.

Stephens, G.L., 2005: Cloud feedbacks in the climate system: A critical review. *J. Climate*, **18**, 237–273.

Stewart, I.T., D.R. Cayan, and M.D. Dettinger, 2004: Changes in snowmelt runoff timing in western North America under a "business as usual" climate change scenario. *Climatic Change*, **62**, 217–232.

Stott, P.A., and C.E. Forest, 2007: Ensemble climate predictions using climate models and observational constraints. *Philosophical Transactions Royal Society A: Mathematical, Physical, and Engineering Sciences*, **365**(1857), 2029–2052, doi: 10.1098/rsta.2007.2075.

Stott, P.A., et al., 2006: Observational constraints on past attributable warming and predictions of future global warming. *J. Climate*, **19**(13), 3055–3069, doi:10.1175/JCLI3802.1.

Stott, P.A., G.S. Jones, and J.F.B. Mitchell, 2003: Do models underestimate the solar contribution to recent climate change? *J. Climate*, **16**, 4079–4093, doi:10.1175/1520-0442(2003)016!4079: DMUTSCO2.0.CO;2.

Stouffer, R.J., et al., 2006: GFDL's CM2 global coupled climate models. Part IV: Idealized climate response. *J. Climate*, **19**, 723–740.

Stouffer, R.J., J. Russell, and M.J. Spelman, 2006: Importance of oceanic heat uptake in transient climate change. *Geophysical Research Letters*, **33**(17), L17704, doi:10.1029/2006GL027242.

Stouffer, R.J., G. Hegerl, and S. Tett, 2000: A comparison of surface air temperature variability in three 1000-yr coupled ocean-atmosphere model integrations. *J. Climate*, **13**, 513–537.

Strack, J.E., R.A. Pielke, and J. Adegoke, 2003: Sensitivity of model-generated daytime surface heat fluxes over snow to land-cover changes. *J. Hydrometeorology*, **4**, 24–42.

Stratton, R.A., 1999: A high resolution AMIP integration using the Hadley Centre model HadAM2b. *Climate Dynamics*, **15**, 9–28.

Sturm, M., et al., 2005: Changing snow and shrub conditions affect albedo with global implications. *J. Geophysical Research–Biogeosciences*, **110**, Art. No. G01004.

Sturm, M., et al., 2001: Snow-shrub interactions in Arctic tundra: A hypothesis with climatic implications. *J. Climate*, **14**, 336–344.

Sud, Y.C., and G.K. Walker, 1999: Microphysics of clouds with the Relaxed Arakawa–Schubert Scheme (McRAS). Part II: Implementation and performance in GEOS II GCM. *J. Atmospheric Sciences*, 56(18), 3221–3240.

Sui, C.-H., X. Li, and K.-M. Lau, 1998: Radiative-convective processes in simulated diurnal variations of tropical oceanic convection. *J. Atmospheric Sciences*, **55**, 2345–2359.

Sun, S., and J. Hansen, 2003: Climate simulations for 1951–2050 with a coupled atmosphere-ocean model. *J. Climate*, **16**, 2807–2826, doi:10.1175/1520–0442.

Sun, S., and R. Bleck, 2001: Atlantic thermohaline circulation and its response to increasing CO_2 in a coupled atmosphere-ocean model. *Geophysical Research Letters*, **28**, 4223–4226.

Sun, Y., et al., 2006: How often does it rain? *J. Climate*, **19**, 916–934.

Svensmark, H., 2007: Cosmoclimatology: A new theory emerges. *Astronomy Geophysics*, **48**, 118124, doi:10.1111/j.1468-4004.2007.48118.x.

Takle, E.S., et al., 1999: Project to Intercompare Regional Climate Simulations (PIRCS): Description and initial results. *J. Geophysical Research*, **104**(D16), 19443–19461.

Talley, L.D., J.L. Raid, and P.E. Robbins, 2003: Data-based meridional overturning stream functions for the global ocean. *J. Climate*, **16**, 3213–3226.

Tao, W.-K., 2007: Cloud-resolving modeling. *J. Meteorological Society Japan*. 125th Anniversary Special Issue, **85B**, 305-330.

Tao, W.-K., et al., 2004: Atmospheric energy budget and large-scale precipitation efficiency of convective systems during TOGA COARE, GATE, SCSMEX and ARM: Cloud-resolving model simulations. *J. Atmospheric Sciences*, **61**, 2405–2423.

Tao, W.-K., 2003: Goddard Cumulus Ensemble (GCE) model: Application for understanding precipitation processes. Cloud systems, hurricanes, and the Tropical Rainfall Measuring Mission (TRMM): A Tribute to Dr. Joanne Simpson, Meteorological Monograph. *Bulletin American Meteorological Society*, **51**, 107–138.

Tao, W.-K., et al., 1999: On equilibrium states simulated by cloud-resolving models. *J. Atmospheric Sciences*, **56**, 3128–3139.

Tebaldi, C., et al., 2006: Going to the extremes: An intercomparison of model-simulated historical and future changes in extreme events. *Climate Change* **79**(3–4), 185–211.

Tebaldi, C., et al., 2005: Quantifying uncertainty in projections of regional climate change: A Bayesian approach to the analysis of multimodel ensembles. *J. Climate*, **18**, 1524–1540.

Tenhunen, J.D., et al., 1999: Ecosystem Studies, Land Use, and Resource Management, 1–19 in *Integrating Hydrology, Ecosystem Dynamics, and Biochemistry in Complex Landscapes*, ed. J.D. Tenhunen and P. Kabat, Wiley, Chichester.

Thompson, D.W.J., and S. Solomon, 2002: Interpretation of recent Southern Hemisphere climate change. *Science*, **296**, 895–899, doi:10.1126/ science.1069270.

Thompson, D.W.J., and J.M. Wallace, 2000: Annual modes in the extratropical circulation. Part I: Month-to-month variability. *J. Climate*, **13**, 1000–1016.

Thompson, D.W.J., and J.M. Wallace, 1998: The Arctic Oscillation signature in the wintertime geopotential height and temperature fields. *Geophysical Research Letters*, **25**, 1297–1300.

Thompson, S., et al., 2004: Quantifying the effects of CO_2-fertilized vegetation on future global climate and carbon dynamics. *Geophysical Research Letters*, **31**(23), L23211.

Thomson, A.M., et al., 2005: Climate change impacts for the conterminous USA, Part 3: Dryland production of grain and forage crops. *Climatic Change*, **69**, 43–65.

Thorne, P.W., et al., 2007: Tropical vertical temperature trends: A real discrepancy? *Geophysical Research Letters*, **34**, L16702, doi:10.1029/2007GL029875.

Tiedtke, M., 1993: Representation of clouds in large-scale models. *Monthly Weather Review*, **121**, 3040–3061.

Tiedtke, M., 1989: A comprehensive mass flux scheme for cumulus parameterization in large scale models. *Monthly Weather Review*, **117**, 1779–1800.

Tjernström, M., et al., 2005: Modelling the Arctic boundary layer: An evaluation of six ARCMIP regional-scale models with data from the SHEBA project. *Boundary-Layer Meteorology*, **117**, 337–381.

Tjernström, M., M. Zagar, and G. Svensson, 2004: Model simulations of the Arctic atmospheric boundary layer from the SHEBA year. *Ambio*, **33**, 221–227.

Tjoelker, M.G., J. Oleksyn, and P.B. Reich, 2001: Modelling respiration of vegetation: Evidence for a general temperature-dependent Q10. *Global Change Biology*, **7**(2), 223–230.

Tompkins, A., 2002: A prognostic parameterization for the subgrid-scale variability of water vapor and clouds in large-scale models and its use to diagnose cloud cover. *J. Atmospheric Sciences*, **59**, 1917–1942.

Trenberth, K.E., J. Fasullo, and L. Smith, 2005: Trends and variability in column-integrated atmospheric water vapor. *Climate Dynamics*, **24** (7–8), 741–758, doi:10.1007/s00382-005-0017-4.

Trenberth, K.E., et al., 1998: Progress during TOGA in understanding and modeling global teleconnections associated with tropical sea surface temperatures. *J. Geophysical Research*, **103** (special TOGA issue), 14291–14324.

Trenberth, K.E., and T.J. Hoar, 1997: El Niño and climate change. *Geophysical Research Letters*, **24**, 3057–3060.

Trenberth, K.E., and J. Hurrell, 1994: Decadal atmosphere-ocean variations in the Pacific. *Climate Dynamics*, **9**, 303–319.

Trier, S. B., et al., 1996: Structure and evolution of the 22 February 1993 TOGA COARE squall line: Numerical simulations. *J. Atmospheric Sciences*, **53**, 2861–2886.

Tripoli, G.J., 1992: A nonhydrostatic mesoscale model designed to simulate scale interaction. *Monthly Weather Review*, **120**, 1342–1359.

Tripoli, G.J., and W.R. Cotton, 1989: Numerical study of an observed orogenic mesoscale convective system. Part 2: Analysis of governing dynamics. *Monthly Weather Review*, **117**, 305–328.

Troccoli, A., and T.N. Palmer, 2007: Ensemble decadal predictions from analysed initial conditions. *Philosophical Transactions Royal Society A*, **365** (No. 1857).

Tselioudis, G., and C. Jakob, 2002: Evaluation of midlatitude cloud properties in a weather and a climate model: Dependence on dynamic regime and spatial resolution. *J. Geophysical Research*, **107**, 4781.

Tselioudis, G., Y.-C. Zhang, and W.R. Rossow, 2000: Cloud and radiation variations associated with northern midlatitude low and high sea level pressure regimes. *J. Climate*, **13**, 312–327, doi:10.1175/1520-0442(2000).

Tsushima, Y., A. Abe-Ouchi, and S. Manabe, 2005: Radiative damping of annual variation in global mean surface temperature: Comparison between observed and simulated feedback. *Climate Dynamics*, **24**, 591–597.

Twomey, S., 1977: The influence of pollution on the short wave albedo of clouds. *J. Atmospheric Sciences*, **34**, 1149–1152.

Ueda, H., et al., 2006: Impact of anthropogenic forcing on the Asian summer monsoon as simulated by eight GCMs. *Geophysical Research Letters*, **33**, doi:10.1029/2005GL025336.

Uotila, P., et al., 2007: Changes in Antarctic net precipitation in the 21st Century based on IPCC model scenarios. *J. Geophysical Research*, **112**, D10107, doi:10.1029/2006JD007482.

Uppala, S.M., et al., 2005: The ERA-40 reanalysis. *Quarterly J. Royal Meteorological Society*, **131**(612), 2961–3012, doi:10.1256/qj.04.176. Index is at www.ecmwf.int/research/era/ ERA-40_Atlas/docs/index.html.

van Oldenborgh, G.J., S.Y. Philip, and M. Collins, 2005: El Niño in a changing climate: A multi-model study. *Ocean Science*, **1**, 81–95.

van Ulden, A.P., and G.J. van Oldenborgh, 2006: Large-scale atmospheric circulation biases and changes in global climate model simulations and their importance for climate change in Central Europe. Atmospheric Chemistry Physics, **6**(4), 863–881

VanRheenen, N.T., et al., 2004: Potential implications of PCM climate change scenarios for Sacramento–San Joaquin River basin hydrology and water resources. *Climatic Change*, **62**, 257–281.

Vavrus, S., et al., 2006: The behavior of extreme cold air outbreaks under greenhouse warming. *International J. Climatology*, **26**, 1133–1147.

Velicogna, I., and J. Wahr, 2006: Acceleration of Greenland ice mass loss in spring 2004. *Nature*, **443**(7109), 329–331.

Vidale, P.L., et al., 2003: Predictability and uncertainty in a regional climate model. *J. Geophysical Research*, **108**(D18), 4586.

Vinnikov, K.Y., D.J. Cavalieri, and C.L. Parkinson, 2006: A model assessment of satellite observed trends in polar sea ice extents. *Geophysical Research Letters*, **33**, L05704.

Vitart, F., and J.L. Anderson, 2001: Sensitivity of Atlantic tropical storm frequency to ENSO and interdecadal variability of SSTs in an ensemble of AGCM integrations. *J. Climate*, **14**(4), 533–545.

Völker, C., D.W.R. Wallace, and D.A. Wolf-Gladrow, 2002: On the role of heat fluxes in the uptake of anthropogenic carbon in the North Atlantic. *Global Biogeochemical Cycles*, **16**(4), 1138, doi:10.1029/2002GB001897.

Von Storch, H., E. Zorita, and U. Cubasch, 1993: Downscaling of global climate change estimates to regional scales: An application to Iberian rainfall in wintertime. *J. Climate*, **6**, 1161–1171.

Wang, B., 1995: Interdecadal changes in El Niño onset in the last four decades. *J. Climate*, **8**, 267–284.

Wang, C., 2005: A modeling study of the response of tropical deep convection to the increase of cloud condensation nuclei concentration: 1. Dynamics and microphysics. *J. Geophysical Research*, **110**, D21211.

Wang, H., and K.-M. Lau, 2006: Atmospheric hydrological cycle in the tropics in Twentieth Century coupled climate simulations. *J. Climate*, **26**, 655–678.

Wang, M., et al., 2007: Intrinsic versus forced variation in coupled climate model simulations over the Arctic during the Twentieth Century. *J. Climate*, **20**(6), 1093–1107.

Wang, W., and M.E. Schlesinger, 1999: The dependence on convection parameterization of the tropical intraseasonal oscillation simulated by the UIUC 11-layer atmospheric GCM. *J. Climate*, **12**(5), 1423–1457.

Wang, Y., et al., 2004: Regional climate modeling: Progress, challenges, and prospects. *J. Meteorological Society Japan*, **82**, 1599–1628.

Warrach, K., H.T. Mengelkamp, and E. Raschke, 2001: Treatment of frozen soil and snow cover in the land surface model SEWAB. *Theoretical Applied Climatology*, **69**, 23–37.

Webb, M.J., et al., 2006: On the contribution of local feedback mechanisms to the range of climate sensitivity in two GCM ensembles. *Climate Dynamics*, **27**(1), 17–38, doi:10.1007/s00382-006-0111-2.

Webb, M.J., et al., 2001: Combining ERBE and ISCCP data to assess clouds in the Hadley Centre, ECMWF and LMD atmospheric climate models. *Climate Dynamics*, **17**, 905–922.

Webster, P.J., et al., 1998: Monsoons: Processes, predictability, and the prospects for prediction. *J. Geophysical Research*, **103**(C7), 14451–14510, doi:10.1029/97JC02719.

Wei, H., et al., 2002: Calibration and validation of a regional climate model for pan-Arctic hydrologic simulation. *J. Climate*, **15**, 3222–3236.

Wentz, F.J., et al., 2007: How much more rain will global warming bring? *Science*, **317**, 233–235.

Wetherald, R.T., and S. Manabe, 1988: Cloud feedback processes in general circulation models. *J. Atmospheric Sciences*, **45**, 1397–1415.

Whitman, S., et al., 1997: Mortality in Chicago attributed to the July 1995 heat wave. *American J. Public Health*, **87**, 1515–1518.

Widman, M., C.S. Bretherton, and E.P. Salathe, Jr., 2003: Statistical precipitation downscaling over the Northwestern United States using numerically simulated precipitation as a predictor. *J. Climate*, **16**, 799–816.

Wigley, T.M.L., et al., 2005: Effect of climate sensitivity on the response to volcanic forcing. *J. Geophysical Research*, **110**, D09107.1–D09107.8, doi:10.1029/2004JD005557.

Wigley, T.M.L., and M.E. Schlesinger, 1985: Analytical solution for the effect of increasing CO_2 on global mean temperature. *Nature*, **315**, 649–652.

Wilby, R.L., et al., 2004: *Guidelines for Use of Climate Scenarios Developed from Statistical Downscaling Methods*. IPCC Data Distribution Centre, University of East Anglia, U.K., 27 pp. (ipcc-ddc.cru.uea.ac.uk/guidelines).

Wilby, R.L., R. Dawson, and E.M. Barrow, 2002: SDSM: A decision support tool for the assessment of regional climate change assessments. *Environmental Modelling Software*, **17**, 145–157.

Wilby, R.L., et al., 2000: Hydrological responses to dynamically and statistically downscaled general circulation model output. *Geophysical Research Letters*, **27**, 1199–1202.

Wilby, R.L., et al., 1998: Statistical downscaling of general circulation model output: A comparison of methods. *Water Resources Research*, **34**, 2995–3008.

Wilby, R.L., and T.M.L. Wigley, 1997: Downscaling general circulation model output: A review of methods and limitations. *Progress Physical Geography*, **21**, 530–548.

Wilks, D.S., and R.L. Wilby, 1999: The weather generation game: A review of stochastic weather models. *Progress Physical Geography*, **23**, 329–357.

Williams, K.D., et al., 2006: Evaluation of a component of the cloud response to climate change in an intercomparison of climate models. *Climate Dynamics*, **145**, 145–165.

Williams, K.D., M.A. Ringer, and C.A. Senior, 2003: Evaluating the cloud response to climate change and current climate variability. *Climate Dynamics*, **20**, 705–721.

Willson, R.C., and A.V. Mordvinov, 2003: Secular total solar irradiance trend during solar cycles 21–23. *Geophysical Research Letters*, **30**(5), 1199–1202, doi:10.1029/2002GL016038, 2003.

Wilson, D.R., and S.P. Ballard, 1999: A microphysics-based precipitation scheme for the U.K. Meteorological Office numerical weather prediction model. *International J. Climatology*, **125**, 1607–1636.

Wilson, M.F., et al., 1987: Sensitivity of the Biosphere Atmosphere Transfer Scheme (BATS) to the inclusion of variable soil characteristics. *J. Climate Applied Meteorology*, **26**, 341–362.

Wilson, T.B., et al., 2003: Evaluation of the importance of Lagrangian canopy turbulence formulations in a soil-plant-atmosphere model. *Agricultural Forest Meteorology*, **115**, 51–69.

Winkler, J.A., et al., 2002: Possible impacts of projected temperature change on commercial fruit production in the Great Lakes Region. *J. Great Lakes Research*, **28**, 608–625.

Winton, M., 2000: A reformulated three-layer sea ice model. *J. Atmospheric Oceanic Technology*, **17**, 525–531.

Wittenberg, A.T., et al., 2006: GFDL's CM2 global coupled climate models. Part III: Tropical Pacific climate and ENSO. *J. Climate*, **19**, 698–722.

Wood, A.W., et al., 2004: Hydrological implications of dynamical and statistical approaches to downscaling climate model outputs. *Climate Change*, **62**, 189–216.

Woodward, F.I., and M.R. Lomas, 2004: Vegetation dynamics—simulating responses to climatic changes. *Biological Reviews* **79**, 643–670.

Wu, X., et al., 2007: Coupling of convective momentum transport with convective heating in global climate simulations. *J. Atmospheric Sciences*, **64**(4), 1334–1349.

Wu, X., and M.W. Moncrieff, 2001: Long-term behavior of cloud systems in TOGA COARE and their interactions with radiative and surface processes. Part III: Effects on the energy budget and SST. *J. Atmospheric Sciences*, **58**, 1155–1168.

Wyant, M.C., et al., 2006: A comparison of tropical cloud properties and responses in GCMs using mid-tropospheric vertical velocity. *Climate Dynamics*, **27**, 261–279.

Wyant, M.C., M. Khairoutdinov, and C.S. Bretherton, 2006: Climate sensitivity and cloud response of a GCM with a superparameterization. *Geophysical Research Letters*, **33**, L06714.

Wyrtki, K., 1975: El Niño—The dynamic response of the equatorial Pacific Ocean to atmospheric forcing. *J. Physical Oceanography*, **5**, 572–584.

Xie, P., and P.A. Arkin, 1997: Global precipitation: A 17-year monthly analysis based on gauge observations, satellite estimates, and numerical model outputs. *Bulletin American Meteorological Society* **78**, 2539–2558.

Xie, S., et al., 2005: Simulations of midlatitude frontal clouds by single-column and cloud-resolving models during the Atmospheric Radiation Measurement March 2000 cloud intensive operational period. *J. Geophysical Research*, **110**, D15S03.

Xie, S., et al., 2004: Impact of a revised convective triggering mechanism on Community Atmosphere Model, Version 2, simulations: Results from short-range weather forecasts. *J. Geophysical Research*, **109**, D14102, doi:10.1029/2004JD004692.

Xu, K.-M., et al., 2005: Modeling springtime shallow frontal clouds with cloud-resolving and single-column models. *J. Geophysical Research*, **110**, D15S04, doi:10.1029/2004JD005153.

Xu, K.-M., and D.A. Randall, 1998: Influence of large-scale advective cooling and moistening effects on the quasi-equilibrium behavior of explicitly simulated cumulus ensembles. *J. Atmospheric Sciences*, **55**, 896–909.

Xue, Y., et al., 2001: The impact of land surface processes on simulations of the U.S. hydrological cycle: A case study of the 1993 flood using the SSiB land surface model in the NCEP ETA regional model. *Monthly Weather Review*, **129**(12), 2833–2860.

Yamaguchi, K., A. Noda, and A. Kitoh, 2005: The changes in permafrost induced by greenhouse warming: A numerical study applying multiple-layer ground model. *J. Meteorological Society Japan*, **83**, 799–815.

Yang, Z.W., and R. Arritt, 2002: Tests of a perturbed physics ensemble approach for regional climate modeling. *J. Climate*, **15**, 2881–2896.

Yao, M.-S., and A.D. Del Genio, 2002: Effects of cloud parameterization on the simulation climate changes in the GISS GCM. Part II: Sea surface temperature and cloud feedbacks. *J. Climate*, **15**, 2491–2503.

Yeh, P.J.-F., and E.A.B. Eltahir, 2005: Representation of water table dynamics in a land surface scheme. Part 1: Model development. *J. Climate*, **18**, 1861–1880.

Yokohata, T, et al., 2005: Climate response to volcanic forcing: Validation of climate sensitivity of a coupled atmosphere-ocean general circulation model. *Geophysical Research Letters*, **32**(21), L21710.1–L21710.4.

York, J.P., et al., 2002: Putting aquifers into atmospheric simulation models: An example from the Mill Creek Watershed, northeastern Kansas. *Advances Water Resources*, **25**, 221–238.

Yu, H., et al., 2006: A review of measurement-based assessment of aerosol direct radiative effect and forcing. *Atmospheric Chemistry Physics*, **6**, 613–666.

Yu, X., and M.J. McPhaden, 1999: Seasonal variability in the equatorial Pacific. *J. Physical Oceanography*, **29**, 925–947.

Zebiak, S.E., and M.A. Cane, 1987: A model El Niño-Southern Oscillation. *Monthly Weather Review*, **115**, 2262–2278.

Zhang, C., et al., 2006: Simulations of the Madden-Julian Oscillation in four pairs of coupled and uncoupled global models. *Climate Dynamics* **27**(6), 573–592.

Zhang, D., and M.J. McPhaden, 2006: Decadal variability of the shallow Pacific meridional overturning circulation: Relation to tropical sea surface temperatures in observations and climate change models. *Ocean Modelling*, **15**, 250–273.

Zhang, G.J., and N.A. McFarlane, 1995: Sensitivity climate simulations to the parameterization of cumulus convection in the Canadian Climate Centre general circulation model. *Atmosphere–Ocean*, **33**, 407–446.

Zhang, J., and D. Rothrock, 2000: Modeling Arctic sea ice with an efficient plastic solution, *J. Geophysical Research*, **105**, 3325–3338.

Zhang, M.H., et al., 2005: Comparing clouds and their seasonal variations in 10 atmospheric general circulation models with satellite measurements. *J. Geophysical Research*, **110**, D15S02, doi:10.1029/2004JD005021.

Zhang, M., 2004: Cloud-climate feedback: How much do we know? In Observation, Theory, and Modeling of Atmospheric Variability. World Scientific Series on Meteorology of East Asia, Vol. 3, ed. Zhu et al. World Scientific Publishing Co., Singapore, 632 pp.

Zhang, M.H., et al., 2001: Objective analysis of the ARM IOP data: Method and sensitivity. *Monthly Weather Review*, **129**, 295–311.

Zhang, M.H., et al., 1994: Diagnostic study of climate feedback processes in atmospheric general circulation models. *J. Geophysical Research*, **99**, 5525–5537.

Zhang, R., and T.L. Delworth, 2006: Impact of Atlantic multidecadal oscillations on India/Sahel rainfall and Atlantic hurricanes. *Geophysical Research Letters*, **33**, L17712, doi:10.1029/2006GL026267.

Zhang, X., and J.E. Walsh, 2006: Toward a seasonally ice-covered Arctic Ocean: Scenarios from the IPCC AR4 model simulations. *J. Climate*, **19**, 1730–1747.

Zhang, X.-C., 2005: Spatial downscaling of global climate model output for site-specific assessment of crop production and soil erosion. *Agricultural Forest Meteorology*, **135**, 215–229.

Zhang, Y.C., A.N. Huang, and X.S. Zhu, 2006: Parameterization of the thermal impacts of sub-grid orography on numerical modeling of the surface energy budget over East Asia. *Theoretical Applied Climatology*, **86**, 201–214.

Zhu, J., and X.-Z. Liang, 2007: Regional climate model simulations of U.S. precipitation and surface air temperature during 1982–2002: Interannual variation. *J. Climate*, **20**(2), 218–232.

Zorita, E., and H. von Storch, 1999: The analog method as a simple statistical downscaling technique: Comparison with more complicated methods. *J. Climate*, **12**, 2474–2489.

CONTACT INFORMATION

Global Change Research Information Office
c/o Climate Change Science Program Office
1717 Pennsylvania Avenue, NW, Suite 250
Washington, DC 20006
202-223-6262 (voice)
202-223-3065 (fax)

The Climate Change Science Program incorporates the U.S. Global Change Research Program and the Climate Change Research Initiative.

To obtain a copy of this document, place an order at the Global Change Research Information Office (GCRIO) web site: http://www.gcrio.org/orders

---

CLIMATE CHANGE SCIENCE PROGRAM AND THE SUBCOMMITTEE ON GLOBAL CHANGE RESEARCH

William J. Brennan, Chair
Department of Commerce
National Oceanic and Atmospheric Administration
Acting Director, Climate Change Science Program

Jack Kaye, Vice Chair
National Aeronautics and Space Administration

Allen Dearry
Department of Health and Human Services

Anna Palmisano
Department of Energy

Mary Glackin
National Oceanic and Atmospheric Administration

Patricia Gruber
Department of Defense

William Hohenstein
Department of Agriculture

Linda Lawson
Department of Transportation

Mark Myers
U.S. Geological Survey

Jarvis Moyers
National Science Foundation

Patrick Neale
Smithsonian Institution

Jacqueline Schafer
U.S. Agency for International Development

Joel Scheraga
Environmental Protection Agency

Harlan Watson
Department of State

EXECUTIVE OFFICE AND OTHER LIAISONS

George Banks
Council on Environmental Quality

Stuart Levenbach
Office of Management and Budget

Stephen Eule
Department of Energy
Director, Climate Change Technology Program

Howard Frumkin
Centers for Disease Control and Prevention

Katharine Gebbie
National Institute of Standards and Technology

Margaret R. McCalla
Office of the Federal Coordinator for Meteorology

Gene Whitney
Office of Science and Technology Policy